Y. Chupeau · M. Caboche · Y. Henry (Eds.)

Androgenesis and Haploid Plants

Springer
Berlin
Heidelberg
New York
Barcelona
Budapest
Hong Kong
London
Milan
Paris
Santa Clara
Singapore
Tokyo

Y. Chupeau M. Caboche Y. Henry (Eds.)

Androgenesis and Haploid Plants

With 20 Figures and 7 Tables

In Memory of Jean-Pierre Bourgin

INRA EDITIONS

Springer

Dr. Yves Chupeau
Dr. Michel Caboche
Laboratoire de Biologie Cellulaire
INRA Centre de Versailles
78026 Versailles Cedex
France

Dr. Yves Henry
Laboratoire de Biologie des Plantes
Université Paris Sud
IBP Bât. 630
91405 Orsay Cedex
France

ISBN 3-540-64025-8 Springer-Verlag Berlin Heidelberg New York

Library of Congress Cataloging-in-Publication Data

Androgenesis and haploid plants/Chupeau Yves, Michel Caboche, Yves Henry (eds.) p.cm.
„These proceedings evolved from a small meeting . . . at the National Institute for Agronomic Research in Versailles, France, on June 21, 1996"–Pref.
Includes bibliographical references and index. ISBN 3-540-64025-8 (hardcover) 1. Androgenesis in plants – Congresses. 2. Haploidy – Congresses. I. Chupeau, Yves, 1944–. II. Caboche, Michel, 1946–. III. Henry, Yves, 1946–. SB 123.6.A54 1998 631.5'2–dc21 98-14360 CIP

Cover design: design & production GmbH, Heidelberg
Cover photograph: © D. Twell, Mature pollen from gametophytic mutants of *Arabidopsis*
Typesetting: Camera ready by the editors
SPIN 10664628 31/3137 5 4 3 2 1 0 - Printed on acid-free paper

Preface

These proceedings evolved from a small meeting of the French section of the IAPTC (International Association for Plant Tissue Culture) held in memory of Jean-Pierre Bourgin, at the National Institute for Agronomic Research in Versailles, France, on June 21 1996.

The first androgenetic plants were described thirty years ago, when Jean-Pierre Bourgin and Jean-Paul Nitsch demonstrated that normal but haploid plants could develop from pollen grains through in vitro culture of anthers.

Indeed, the interest of breeders for rapid creation of homozygous lines led to a considerable amount of work dealing with the production of haploid and doubled haploid plants.

However, recent achievements in the control of androgenesis, mainly for rapeseed and tobacco pollen grains, have progressively made androgenesis a true scientific tool for modern plant biology. Pollen is the only tissue in higher plants which consists of an isolated cell without cell walls, and with a peculiar cell cycle from meiosis to the asymmetric division of the polarized microspore.

The rapid development of the genetic approach has already led to the isolation of several classes of gametophytic mutants affecting viability, polarity and generating mechanisms, as well as cell fate determinants.

Since it is now technically possible to begin the study of cell fate determination, this study could prove to be particularly useful for differentiated cells such as gametes, which can be experimentally reoriented to sporophytic development towards direct embryo formation.

The editors

Contents

Jean-Pierre Bourgin in a green house of the Center, INRA Versailles, in 1993.

1 The Very First Androgenetic Plants: The Scientific Achievements of Jean-Pierre Bourgin, 1944-1994

Y. Chupeau
Laboratoire de Biologie Cellulaire, INRA Centre de Versailles, F 78026 Versailles, France

Jean-Pierre Bourgin, one of the founders of modern plant cell biology, died on 28 October 1994 whilst enthusiastically involved in refining and optimizing the organization of the laboratory of plant cell and molecular biology (Institut National de la Recherche agronomique, Versailles, France).

Deeply involved in plant tissue culture throughout his twenty eight years of scientific activity, Jean-Pierre was associated with two turning points in this field. After his early demonstration that regeneration of haploid plants was feasible, he made use of haploid cells studying cell genetics.

Considering the history of plant tissue culture, it is striking to realize how the majority of scientists were deeply involved in the search for conditions for inorganized and indefinite growth. The quest for a full experimental verification of the cellular theory, the intellectual incentive in the early attempts, had become a sort of obsession.

However, after the first turning point, the discovery of cytokinins (Miller et al. 1955), around the 1960s some scientists were already convinced of the potential of tissue culture for agronomic applications. In France, this was the case for two major figures, Jean-Paul Nitsch at the CNRS in Gif sur Yvette and Georges Morel at INRA in Versailles. Both were working towards the idea that tissue culture could solve agronomic problems. Both were brightly intelligent and had no difficulty convincing Jean-Pierre Bourgin, who thus appeared as their spiritual son.

After training in agronomic and biological sciences, Jean-Pierre Bourgin was welcome by Jean-Paul Nitsch, in a CNRS laboratory of the "Phytotron complex" at Gif sur Yvette, for his very first experience of laboratory work in 1966. At that time, plant tissue culture had become an exciting field of research since plant regeneration from undifferentiated calli had been observed in two different plant species, namely carrot and tobacco.

In the late 1950s, the characterization of a new class of plant growth substances: the cytokinins, led Skoog and Miller (1957) to the chemical control of the formation of buds on tobacco tissues. At about the same period, in three different laboratories (Germany, USA, Japan) somatic embryogenesis was also described for carrot tissues cultivated *in vitro*. This phenomenon appeared even more astonishing than bud formation, as it was repeatedly possible to induce the formation of embryos out of somatic cells, thus without going through gametogenesis and fertilization.

Furthermore, in the mid 1960s, in India Guha and Maheswari (1966) revealed that pollen grains of *Datura* could also develop into embryos upon *in vitro* culture of anthers. However, in those days the essential goal of tissue culture was still almost purely conceptual: still aimed at the experimental verification of the cellular theory assuming that individual cells are able to divide indefinitely... and that this capacity could be extended to pollen grains.

1.1 Haploid plants

The major interests of Jean-Paul Nitsch were related to the action of growth substances and he made use of tissue culture to analyze their implication in crucial stages of plant development such as flower and fruit formation. Indeed, the role of auxins and cytokinins in bud or embryo formation became major axes of research in Nitsch's laboratory, where Jean-Pierre Bourgin was in charge of the adaptation of carrot tissue culture for induction of somatic embryos. In a few months, he became trained not only to induce and detect somatic embryos but also to culture them through different steps to entire and normal carrot plants.

Then, Jean-Paul Nitsch asked him to verify embryo formation from pollen grains. As the intention was clearly to assess the possibility of obtaining entire haploid plants from a perspective of practical uses, they decided to culture anthers of plants for which processes of regeneration were already well known. Anthers from carrots are not easy to work with, so anthers from various *Nicotiana* were chosen.

By methodically culturing anthers at different stages of development, Jean-Pierre was rapidly and happily able to recognize, inside some anthers in culture, some typical embryo forms. Surprisingly, some of these embryos directly germinated and emerged from the browning anther remains and, after transplantation for a further period of development in vitro, the plantlets were transferred to the greenhouse. A set of regenerated tobacco plants, confirmed as haploid, grew almost normally though , had narrow leaves and sterile flowers.

They soon verified that diploidization could restore the fertility of the plant and, at the same time, very rapidly provide true homozygous plants...

The first paper issuing from this work (Bourgin and Nitsch, 1967) will be one of the most cited in plant biology for the next twenty five years. Indeed, the technique has been extended to many other plant species both dicots and monocots, and plant breeders are progressively more and more engaged in attempts to use homozygous plants in selection schemes.

The first experimental work of Jean-Pierre Bourgin can be considered as the second turning point of plant tissue culture in that the process of haploidization, and haploid plants, were henceforth considered as true scientific subjects. In fact, the idea of doing genetics with plant cells was no longer pure speculation. In the early 1970s for most scientists in this field, it was no longer a question of verifying the cellular theory but more practically to progress to make use of all its practical consequences.

1.2 Selection for biochemical mutants

By the end of 1967, Jean-Pierre was hired by the Institut National de la Recherche Agronomique at the Station Centrale de Physiologie Végétale in Versailles. There, in the Service de Biochimie Cellulaire headed by Georges Morel, Jean-Pierre practicised meristem culture for vegetative multiplication of *Asparagus*, another important breakthrough of plant tissue culture.

After the period of military services Jean-Pierre and I were back in Morel's laboratory in Versailles. We decided to share our experiences in tissue culture to progress towards plant cell genetics. Georges Morel readily accepted the idea and fully supported our initial aims concerning the search for a model system. Both of us were driven by the desire to apply the concepts of microbial genetics to plants. The general idea was to work with haploid cells in the protoplast state, to make use of all the experimental advantages of protoplasts.

At that time, after the pioneering work of Itaru Takebe who described the divisions of protoplasts from tobacco leaves at the Strasbourg symposium in 1970, the idea to work with haploid leaf cells was shared by interested biologists. Confirmation of this choice came rather rapidly as Jean-Paul Nitsch with Kanji Ohyama (1971) and Itaru Takebe with Georg Melchers (1971) succeeded at about the same period to regenerate plants from protoplasts of haploid tobacco plants. We soon verified the practicability of such cultures. However, like every laboratory in this field we faced the erratic behavior of protoplasts in culture. For complex and obscure reasons preparations of protoplasts were so variable, not only in yields but also in viability, that we could hardly think of using such protoplasts for genetic studies. Protoplasts obtained from plant tissues already cultivated *in vitro* appeared easier to use, and most laboratories made use of this facility. However, it was already clear that *in vitro* cultured plant cells were rapidly accumulating mutations and even large karyotype deviations.

We chose to try to unravel the respective and intricate roles of genotype and growth conditions to secure repeatable isolations of protoplasts directly from plants, through the use of a known genotype. For the same reasons, we paid specific attention to the effects of the maceration conditions on the ultimate fate of protoplasts in culture. In spite of the noticeable lag before the first experimental uses of protoplasts in our hands, we never regretted these early basic studies as they gave a firm basis to most of the developments in somatic genetics in the laboratory, and this remains true nowadays despite the many extensions to the use of protoplasts and genetics...

A first achievement came with a standardization of our conditions for leaf protoplast preparations, making use of a specific haploid clone of *N. tabacum* cv Xanthi, which appeared more effective than other haploids from various tobacco cultivars. This clone not only yielded viable protoplasts, but also higher rates of colony formation and ultimately of bud regeneration. One diploidized line (XHFD8) of this very plant has been continuously used in our laboratory as well as in many others... since 1974. However, we considered this haploid clone more as an intermediate tool, as tobacco is an amphidiploid species. Jean-Pierre pursued his quest for a true haploid model system with different *Nicotiana* (*alata* n=9, *sylvestris* n=12, *plumbaginifolia* n =10). In the late 1970s he proposed *N. plumbaginifolia* as one of the best adapted models. Indeed, this was welcome in different groups engaged in plant cell genetics, and despite the widespread use of *Arabidopsis thaliana* in plant molecular genetics, *N. plumbaginifolia* is still used by numerous teams including some in Versailles.

During the course of this tedious progress in plant protoplast handling, Jean-Pierre began to attempt the different steps towards *in vitro* selection of biochemical mutants. Comparison of various mutagenic treatments made him rapidly rely on U.V. mutagenesis, which he found efficient and practical, as well as being less dangerous in routine uses than chemical mutagens.

It was typical of Jean-Pierre that he proceeded by carefully looking for a model system of selection. He decided to adapt to plant cells the study of the biosynthesis of branched amino-acids. At that time, it was one of the better known systems of regulation from microbial genetics. Jean- Pierre established a method for the isolation of biochemical mutants in tobacco by using toxic doses of valine as a selective agent. He was also able to demonstrate that an excess supply of this amino-acid caused feed-back inhibition of isoleucine and leucine biosynthesis in plants, as has been observed in bacteria. A series of valine-resistant mutants were selected from protoplasts. Plants regenerated from these protoplasts were themselves resistant, and were capable of transmitting this characteristic to their progeny. The publication of this work in 1978, constitutes one of the first examples of the application of microbial genetic concepts to higher eukaryotes. This work resulted in his thesis for the Doctorat d'Etat in Natural Sciences he defended in 1982. Afterwards, this process was extended to *N. plumbaginifolia* and different species, by students of Jean-Pierre as well as other groups.

1.3 The inspired leader

At the very beginning of this research in cell genetics, the laboratory was profoundly distressed by the sudden death of Georges Morel in 1973. In 1976,

INRA offered Jean-Pierre the directorship of the laboratory, from then on called the Laboratoire de Biologie Cellulaire. This decision resulted in the fine body of research which has been produced here ever since.

Under the impetus of Jean-Pierre, profound changes were made in the following years.Thanks to his managerial skills and to his scientific leadership, Jean-Pierre made this laboratory a world-wide renowned centre for plant cell and molecular genetics. From the day of his directorship, he established a welcoming climate of confidence, and mutual respect, and many scientists asked to be part of the team. The newcomers shared his enthusiasm for taking advantage of the new possibilities in plant biology in order to exploit the complementarity between cell genetics and molecular biology for the study of fundamental and agronomic problems. The boom in plant genetic engineering that followed confirmed the validity of these early scientific choices. At the end of the 1980s, the laboratory already numbered more than one hundred people, amongst whom were more than thirty researchers and close to forty graduate students. Scientific breakthroughs mark this evolution, in fields as varied as those of cell and molecular biology (protoplast culture of species of agronomical interest: rapeseed, lettuce, sunflower, chicory...; transfer of cytoplasmic male sterility in rapeseed through somatic hybridization ; several methods of gene transfer...), plant physiology (mechanisms of nitrate assimilation, mode of action of phytohormones in higher plants), plant-pathogen interaction, developmental biology, studies on plant genomes and transposable elements, and plant biotechnology.

These results were made possible in large part thanks to the efforts of Jean-Pierre whose aim was to attract strong scientific personalities and train new talents, to mobilize and coordinate each and everyone's efforts with the concern of permanent efficiency. His colleagues, the students and young scientists who have stayed in the laboratory, and all those who were close to him will keep the memory of an attractive personality, full of communicative energy, with a profound and sincere modesty, a dedication and enthusiasm without limit and an unshakable faith in the virtues of intelligence and communication. Willing to hear all, demanding towards himself and others, he gave a unique style to the position he held.

Overall, Jean-Pierre was able to develop and maintain within the laboratory a particular way of operating and a state of mind, based on union and cooperation, which is still operating today. This was due to his capacity to inspire a spirit of strong and enthusiastic dedication to agriculture, through the constant aim, shared by everyone, of the creation of new tools for plant biology and plant breeding. All of us in the laboratory owe something to his style of activity and human relations. His joyful enthusiasm and the warm deep honesty Jean-Pierre used to communicate are still a source of inspiration in the laboratory.

References

Bourgin JP, Nitsch JP (1967) Obtention de Nicotiana haploides à partir d'étamines cultivées in vitro. Ann Physio Vég 9:377 – 382

Bourgin JP (1978) Valine-resistant plants from in vitro selected tobacco cells. Mol Gen Genet 161:225 - 230

Guha S, Maheshwari Sc (1966) Cell division and differentiation of embryos in the pollen grains of Datura in vitro. Nature 212:97 – 98

Nitsch JP, Ohyama K (1971) Obtention de plantes à partir de protoplastes haploides cultivés in vitro. C R Acad Sci Paris 273:801 - 804

Miller CO, Skoog F, von Saltza MH, Strong FM (1955) Kinetin a cell division factor from desoxyribonucleic acid. J Chem Soc 77:1392

Skoog F, Miller CO (1957) Chemical regulation of growth and organ formation in plant tissues cultured in vitro. Symp Exp Soc Biol 11:118 – 130

Takebe I, Labib G, Melchers G (1971) Regeneration of whole plants from isolated mesophyll protoplasts of tobacco. Naturwissenschaften 58:318 – 320

2 A Critical Analysis of Existing Haploidization Techniques

C. Horlow[1], and C. Raquin[2]
[1]Laboratoire de Génétique et d'Amélioration des Plantes, INRA Centre de Versailles, F 78026 Versailles Cedex, France
[2]Systèmatique et Ecologie Végétales, URA 121 CNRS, Université d'Orsay, F 91405 Orsay Cedex, France

2.1 Introduction

Among the different reported definitions of haploidy those of Chase (1951) seem to be the most convenient. Haploidy refers to any organism, tissue or cell, having the chromosomal constitution similar to the normal gamete of a given species. Haploid plants which originate from a gametic cell in the embryo sac (gynogenesis) or in the pollen (androgenesis) correspond to this definition.

2.1.1 Interests

Selection at the haploid level. Many genes expressed at the diploid level are also expressed at the haploid level which makes selection at either level possible (Tanskley et al., 1981; Willing and Mascarenhas, 1984; Mascarenhas, 1989). Selection at the haploid level may be more effective than at the diploid level because recessive alleles are not masked. Haploid selection has been successful in anther culture for example for salt tolerance in barley (Ye et al., 1987), disease resistance in tobacco (Deaton et al., 1992) and wheat coleoptile length/plant height (Bozorgipour and Snape, 1992).

Haploids can be used at the n level for genetic studies and constitute a choice material for cellular mutagenesis. The mutants can be screened under given conditions and immediately express their variation in regenerated plants. As examples of successes, methionine sulphoximine resistant tobacco cellular clones were selected from protoplast suspension of haploid tobaccos (Carlson, 1973). J.P. Bourgin (1978) obtained a mutant tobacco line valine resistant. Kaul (1986) induced male sterile mutants among doubled haploids obtained through anther culture in rice.

After chromosome doubling. Haploids more often are used after chromosome doubling as double haploids (DH). Such double haploids are interesting for breeders for the following three reasons:

Time saving to produce inbred lines. This point was mentioned by Blakeslee and Belling as early as 1924. In maize the time from anther culture to R1 seed could be as little as seven months (Murigneux et al. 1993). A comparison between doubled haploid and conventional plant breeding methods for winter wheat has been made by Baenzinger et al. (1984) and De Buyser et al. (1987). Two to four years could be saved using haploids instead of F5-derived lines.

Strict homozygosity of the double haploid (DH) plants. The doubled haploid lines are homozygous and homogeneous, reported through the two most extensive studies made in maize with molecular markers. Bentolina et al. (1992) studied 72 doubled haploid lines from a cross with 87 RFLPs markers. All the DH lines were homozygous for one of the parental RFLPs. The other study (Murigneux, 1993) used 100 markers on two families of DH lines, one from a synthetic population and the other from two crosses. The overall heterozygosity of the DH lines was evaluated at less than 1% for the loci assayed.

Unshifted screen. There is no recombinational shift due to residual heterozygosity for the agronomical characters observed. Trait values estimated for a DH will be the same in its progeny.

Now, about seventy five years after the first haploid sporophyte in Jimson weed was reported by Blakeslee et al.(1922), what are the available and efficient techniques for haploid induction?

2.1.2 Pathways to haploidy

Fig. 2.1 schematizes the three different main mechanisms which can be involved to divert normal gametogenesis and fertilization toward haploidy, ie: abnormalities of the embryo development, accidents in fertilization and disturbance of gametogenesis by *in vitro* culture of the gametophytes.

Two preliminary comments must be made:

- There is no obligatory concordance between the biological pathway and the technique of haploidization. For an example, abnormalities of embryo development can occur spontaneously or after an interspecific cross. On the other hand an interspecific cross can lead either to a disturbance of normal embryo development *(Hordeum vulgare x Hordeum bulbosum)* or to an accident of fertilization *(Solanum tuberosum x Solanum phureja).*

- None of the methods of haploidization can be applied to all species.

Therefore we choose to classify the different methods of haploidization according to their requirements of *in vitro* culture.

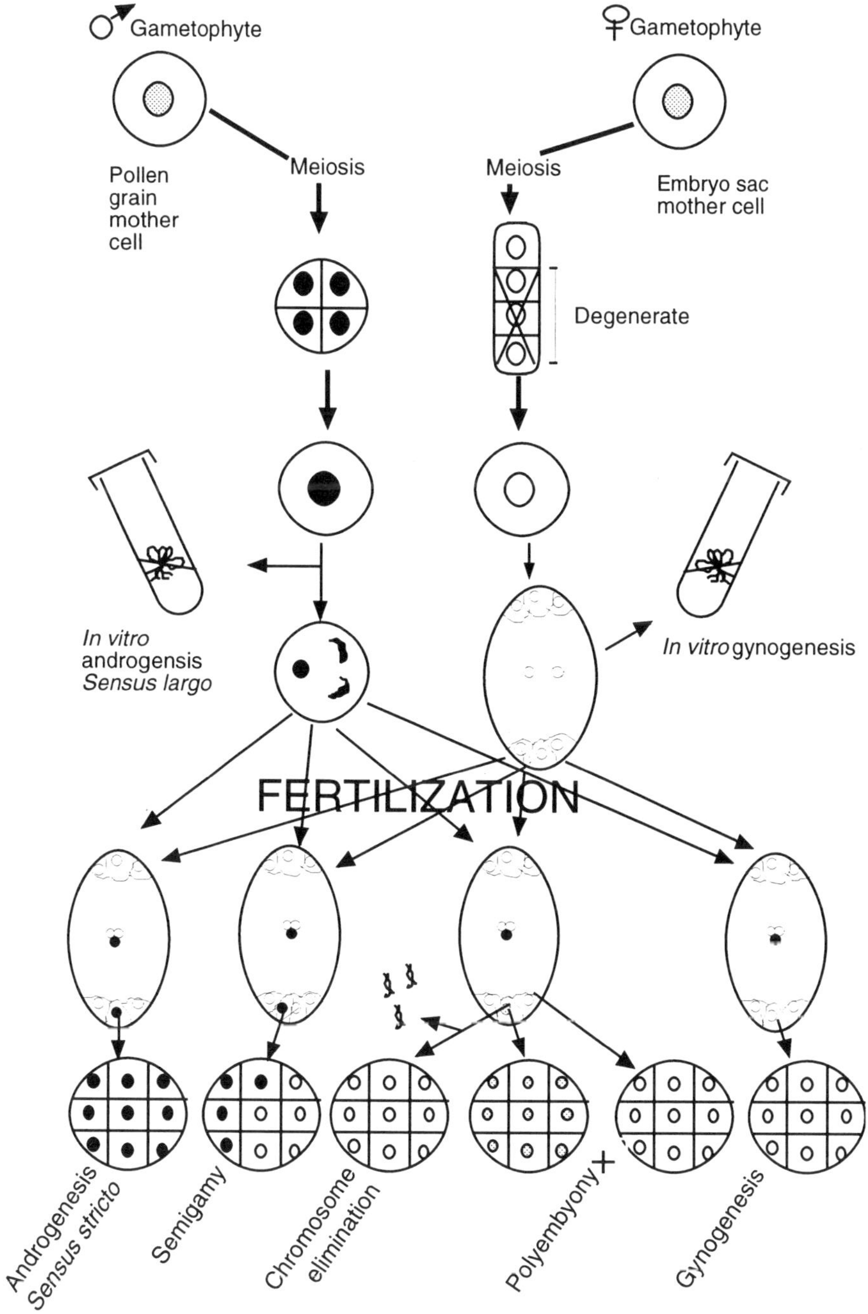

● n male gametic nucleus. ○ n female gametic nucleus. ◎ 2n nucleus.

Fig. 2.1 : n ⇄ 2n cycle and pathway to haploidy.

2.2 Available methods for haploid production

2.2.1 Total requirement: the *in vitro* culture of male or female gametophytes.

Although the first attempt was performed on the megagametophyte of Gymnosperms (La Rue 1948), the success with anther culture of *Datura innoxia* by Guha and Maheshwari (1964, 1966) and anther culture of tobacco by Bourgin and Nitsch (1967) opened the way to in vitro androgenesis which is up to now applied to the largest number of species.

The *in vitro* culture of the male gametophyte (anther or pollen culture). The large development of this method has stimulated many reviews. In 1980 Guha and Maheshwari already reported success in 153 species belonging to 52 genera and 23 families. Recent books were published by Bajaj (1990) and Jain et al. (1996). Therefore our purpose is not to describe the technique of anther or pollen culture.

Two particular directions of research in androgenesis which seem to be the most promising for understanding the mechanisms of pollen embryogenesis and increasing efficiency are currently especially developed in several laboratories:

Genetic control of androgenesis. The genetic control of the different steps of androgenesis has been demonstrated in several species; for a review see Henry et al. (1994).

Physiological and biochemical aspects of the first steps of the androgenetic embryo development. Microspore culture initiated on *Datura* (Nitsch and Norreel 1973) became very efficient in some species such as rapeseed (Lichter 1982), barley (Sunderland and Xu 1982, Hoekstra et al. 1992), tobacco (Nitsch 1974, Heberlee-Bors and Reinert 1979) and permit studies on molecular, biochemical and cytological events during the induction of microspore embryogenesis. These aspects will be developed in the next chapters.

The *in vitro* culture of the female gametophyte (ovary or ovule culture). The first gynogenetic haploid of barley obtained by *in vitro* culture of unpollinated ovaries was reported by San Noeum (1976). A list of the species haploidized by ovary or ovule culture is given in Table 2.1. The protocols for anther or ovary culture are rather similar. The most important difference probably concerns the developmental stage of the gametophyte. Inside the anther as well as outside, androgenetic induction generally requires the culture of uninucleate microspores (cereals and cruciferous) or young pollen grains at or just after the first mitosis for tobacco (Maheshwari et al. 1980). The ovaries or ovules containing nearly mature embryos sacs seem to be the most suitable material for "*in vitro* gynogenesis" (San and Demarly, 1984). The yield of both techniques depends strongly upon the genotype of the donor plant and can be enhanced by physical factors before the *in vitro* culture of gametophytes such as thermal pretreatment (Nitsch and Norreel, 1973) or slight irradiation (Sangwan and Sangwan, 1986).

Table 2.1. In vitro gynogenesis: unfertilized ovary or ovule culture, chronological successes for different species.

Species	References	Culture	Results, ploidy level
Hordeum vulgare	San Noeum (1976)	ovary	E. Pl. n
Triticum aestivum	Zhu and Wu (1979)	ovary	C. Pl. n
Nicotiana rustica	Zhu and Wu (1979)	ovary	E. Pl. n
Oryza sativa	Asselin de Beauville (1980)	ovary	C. E. PL. n-2n
Gerbera jamesonii	Sitbon (1981)	ovule	E. Pl. n
Nicotiana tabacum	Wu and Chen (1982)	ovary	E. Pl. n
Zea mais	Ao et al.(1982)	ovary	E. Pl. n
Lilium davidii	Gu and Cheng (1983)	ovary	E. Pl. n-2n
Beta vulgaris	Hosemans and Bossoutrot (1983)	ovule	E. Pl. n-2n
Petunia axillaris	De Verna and Collins (1984)	ovule	E. Pl. n
Populus simonigra	Wu and Xu (1984)	ovary	E. Pl. n
Heliantus annuus	Cai and Zhou (1984)	ovary	E. Pl. n
Solanum tuberosum	Tao et al.(1985)	ovary	C. Pl. n
Hevea brasiliensis	Chen et al. (1988)	ovule	E. Pl. n-2n
Allium cepa	Campion and Azzimonti (1988)	ovule	E. Pl. n-2n
var. aggregatum	Cohat (1994)	fl. bud	E. Pl. n-2n>2n
Melandrium album	Mol (1992)	ovule	E. Pl. n-2n

C.= callus, E.= embryo, Pl.= plant.

Androgenesis versus gynogenesis: advantages and limits. Up to now the number of species responsive to androgenesis is over ten times higher than those responsive to gynogenesis. Moreover the number of pollen grains and therefore the potential number of haploid plants is much higher than is the number of ovules. Nevertheless over thirty years after the first successes one can consider that only for four species, ie: rapeseed, barley, rice (japonica types) and tobacco (still with difficulties on chromosome doubling), is there no practical limit to the androgenetic haploid production. This number of species appears very low and yield remains too far from the theoretical potential of both techniques to take this limit into account.

Some families such as Compositae, respond better to gynogenesis than to androgenesis and within the poorly responsive family Chenopodiaceae, ovary culture is quite efficient for haploidizing sugarbeet.

Anther or ovary culture produce variable proportions of non haploid plants depending on the species and culture conditions, (for reviews see Rao and Suprasanna, 1996 and Keller and Korzun 1996). Despite some cases of heterozygous diploids issued from unreduced microspores in rye (Wenzel et al. 1976), spontaneous diploids issued from anther culture are generally homozygous. If their frequency is enough, spontaneous DH permit to avoid chromosome doubling which is time and money consuming. For rapeseed spontaneous diploids are common enough to be the only material used by breeders.

Ovary or ovule culture, using a nearly mature embryo sac, run the risk of accidental fertilization prior to culture. This strongly increases the possibility of heterozygous plants amongst the spontaneous diploids and these must be controlled for.

Classifying the number of responsive species, gametophyte culture and especially anther culture remain the most commonly used technique for producing haploids. They are however not universal and other methods can be very suitable. These require more or less *in vitro* culture technique.

2.2.2 Total or partial requirement: gametophyte treatment before fertilization, interspecific hybridization

In such cases embryo rescue is generally necessary if the formation or development of the endosperm is disturbed.

Table 2.2. Radio induced gynogenesis, a list of species haploidized by pollination with irradiated pollen followed by the in vitro embryo rescue.

Species	References	Dose (Gy)	Culture
Hordeum vulgare	Powell et al. (1983)	<20	E (1Pl.)
Triticum aestivum	Snape et al. (1983)	20	E. (1Pl.)
Petunia x hybrida	Raquin (1985)	500-1500	F.
Cucumis melo	Sauton and Dumas de Vaulx (1987)	300	E.
Daucus carota	Rode and Dumas de Vaulx (1987)	300-500	S. (2Pl.)
Malus x domestica	Zhang et al. (1988)	500-1000	S.+E.
Cucumis sativus	Truong-André (1988)	800-1200	E.
Brassica oleracea	Doré (1989)	800-1200	S.+E.
Actinidia deliciosa	Pandey et al. (1990)	500-900	S.
Allium cepa	Dore and Marie (1993)	150	S.
Pyrus communis	Bouvier et al. (1993)	250-500	S.
Rosa x hybrida	Meynet et al. (1994)	500-1500	S.+E.

F.= fruit, S.= seed, E.= embryo, Pl.= plant

Gametophyte treatment before pollination. The ability to produce *in situ* haploids by different methods has been widely exploited. Physical and chemical treatments, generally of the male gametophytes, have been performed to prevent the formation of two functional sperm nuclei, but leaving the pollen capable of germination. Fertilization with this sterile pollen stimulates the parthenogenetic development of an embryo (Illies, 1964).

Data concerning chemical treatments are rare and the results are often not reproducible. One successful example is that of poplar haploids produced after treatment of the pollen with toluidine blue (Illies, 1974).

More data are available concerning physical treatments of pollen. The first haploid *Datura* was observed after cold treatment of the flowers (Blakeslee et al., 1922). Ionizing radiation was used as physical sterilising agent as early as 1933, (for a review see Lacadena, 1974). Despite a large number of attempts the haploid production generally was poor. It seems as if there were two unresolvable constraints. If the irradiation dose was low (< 100 Gy) only few haploids could be found and irradiation increases the mutation rates at markers loci used to sort haploids from diploids. If the dose was high (1000 Gy) seed set occurred but no viable seed could be harvested (Straub, 1973). It is very probable that pollen irradiation induced a deficient endosperm. Therefore embryo rescue was required for the development of the haploid embryos. A list of the species haploidized through pollen irradiation and embryo rescue is given in Table 2.2. However, except in melon and petunia, irradiation of the pollen is not an efficient method for a mass production of haploids. Irradiation of the embryo sac followed by pollination with non irradiated pollen permits the production of *in situ* androgenetics haploids in *Crepis tectorum* (Gerassimova, 1936), *Antirrhinum majus* (Ehrensberger, 1948) and after *in vitro* ovary culture in *Petunia hybrida* and *Petunia parodii* (Raquin et al., 1989).

Interspecific or intergeneric hybridizations. The possibility of the elimination of the chromosomes of one of the partners by interspecific and intergeneric hybridizations has been exploited. For example, a large number of barley haploids have been obtained from crosses between *Hordeum vulgare* (cultivated species) and *Hordeum bulbosum* (wild species) (Symko, 1969; Kasha and Kao, 1970). Regardless of the direction of the cross, the haploids recover the *Hordeum vulgare* genome. Cytological investigations have demonstrated that fertilization is normal but the chromosomes of *Hordeum bulbosum* are eliminated during the first cell divisions during embryo formation (Subrahmanyam and Kasha, 1973; Bennett et al., 1976). The process of chromosome elimination also affects the endosperm, which degenerates. As these embryos are not accompanied by a functional endosperm and could not continue their complete growth on the mother plant, "embryo rescue" by *in vitro* culture is necessary. Other interspecific hybridizations in potato (Hougas et al., 1964), melon (Dumas de Vaulx, 1979) and *Brassica* (Chen and Heenen, 1989) have allowed isolation of haploid plants. By intergeneric crosses between wheat plants used as female parent and maize plants used as pollinators, haploid wheat plants have been obtained after "embryo rescue" (Laurie and Bennett, 1988).

2.2.3 Total independence of the *in vitro* culture: spontaneous haploids

Several phenomena occurring during the reproduction process such as spontaneous parthenogenesis, polyembryony and spontaneous androgenesis can be responsible for the appearance of spontaneous haploids.

Spontaneous parthenogenesis. Spontaneous parthenogenetic haploids occur with a frequency estimated at 1 haploid for 1000 plants after normal pollination in maize (Chase, 1969), and 1 haploid for 1100 plants after normal pollination in tobacco (Burk, 1962). A very high frequency (3.23 %) of maize maternal haploids was obtained when the plants were crossed with the pollen of "stock 6" line, in which for some unknown reason, the sperm nucleus malfunctions during syngamy and is lost (Coe, 1959).

In potato, crosses between the tetraploid cultivated species *Solanum tuberosum*, and some clones of the diploid wild-type species *S. phureja*, constitute a method of producing maternal dihaploids. It was demonstrated in this case, that the sperm nuclei from a proportion of *S. phureja* pollen tubes fuse in a restitution nucleus that fertilizes the central cell of the embryo sac, whereas the egg cell develops into a parthenogenetic embryo. In this case, double fertilization by haploid sperm cells (coming from the normal pollen tubes) leads to the abortion of the endosperm due to the imbalanced genome ratio (x + 4x'). Pratically, because almost all hybrids are lost, this only permits recovery of dihaploids (Hougas et al., 1964).

From the cross *N. africana* Merxm. (as male parent) by *N. tabacum* (as female parent), Burk et al. (1979) found that the majority of the progeny died after germination. Only 0.25% to 1.42% of the seedlings grew normally, and were identified as F1 hybrids or as gynogenetic haploids.

Polyembryony. In polyembryonic seeds, sometimes, one plantlet of twin or triplet seedlings can be haploid. Consequently, polyembryonic seeds, which are easily recognizable, are a source of spontaneous haploids which has often been used in practice, for example in *Asparagus* (Thévenin, 1968). It was demonstrated directly in *Linum usitatissimum* L., by cytological studies, that in a proportion of embryo sacs the initial egg cell divides before fertilization producing two equipotential egg cells, one of which is fertilized whilst the other develops by parthenogenesis (Levieil et al.,1985). The proportion of such additional divisions, is correlated with the frequency of polyembryony, genetically controlled in this species by recessive genes (Kappert, 1933, 1950).

Spontaneous (or *in situ*) androgenesis. Androgenesis is a form of apomixis in which the male gamete develops directly into an embryo in the embryo sac. In *Nicotiana* the first androgenetic haploid was discovered and selected on morphological characters (Kostoff, 1929; Clausen and Lammerts, 1929).

In *Brassica napus* L., spontaneous diploid androgenetic individuals were obtained. The maternal line N°7076 was characterized by a high value of erucic acid in the seeds and white flowers. The cultivars "Topas" and "Puma" (paternal parents) were characterized by zero erucic acid and yellow flowers. From the cross of the line N°7076 x Topas, up to 21% spontaneous androgenetic individuals can be obtained. However the reciprocal crosses yielded no androgenetic individuals. The genetic control of spontaneous androgenesis in this material is not yet elucidated (Chen and Heenen, 1989).

The androgenetic progeny can have a haploid or diploid paternal nuclear constitution. If the male gamete develops in the maternal cytoplasm, a new combination of nucleus and cytoplasm originates (as suggested by Chase, 1951; Goodsell, 1961). This process is of interest for the plant breeder because it leads to the immediate production of alloplasmic stocks (Chase, 1963; Horlow et al., 1993).

The principal limitation of the use of the natural phenomenon of androgenesis for the conversion of fertile lines into male sterile lines is the very low occurency of androgenetic haploids. Chase (1963) In maize the frequency is one androgenetic haploid for 80,000 plants (Chase, 1963). In the same species, the presence of a spontaneous mutation termed indeterminate gametophyte (ig) in the inbred Wisconsin-23 (used as maternal genome) can increase the frequency of spontaneous androgenesis to 3% (Kermicle, 1969).

2.3 Selection of spontaneous haploids by using genetic markers

2.3.1 Dominant marker

A distinction between parthenogenetic and androgenetic origin can be made when the parents differ in respect to one or more characters. The identification of haploidy by morphological appearance can be a tedious and often inaccurate method and the progenies must be cultivated until the adult stage. The dominant marker would be more interesting if its expression appeared very early during plantlet development. In maize, genetic markers identified at the seed or seedling stages have been used to detect maternal haploids (Chase, 1952). CI , the inhibitor allele at the C locus, inhibits anthocyanin pigmentation in both endosperm and scutellum in maize seeds. Coe and Sarkar (1964) have used these genes to screen visually for parthenogenetic haploids with coloured scutella among the colourless seeds obtained from a cross between the maternal parent CC pollinated by the parent CICI. In a reciproqual cross, only the androgenetic haploids were selected (Sarkar, 1972). This technique was easy to use for selecting parthenogenetic and androgenetic haploids in maize but its use was very limited. The necessary markers are not present or desired in all agronomic lines.

In the progeny from crosses between a cytoplasmic male sterile line "Texas" carrying the dominant marker "purple root" and a fertile N6 line devoided of this dominant marker, two plantlets with white roots were isolated by Goodsell (1961). The plants were androgenetic haploids of the line N6 and after chromosome doubling, male sterile cytoplasmic diploid seedlings of the line N6 were obtained.

2.3.2 Dominant lethal marker

If the dominant marker is lethal at the plantlet stage, only the haploids survive and the selection of these individuals is facilitated.

Pelletier et al. (1987) used a single dominant allele, Rac-, conferring to the plants the inability to produce roots. Only by grafting, the plants can survive and give progenies. With this marker, only androgenetic haploid individuals or gynogenetic haploid individuals, depending on whether the homozygous rootless parent used was female or male, survive. This marker was used for the screening of spontaneous haploids and to perform direct cytoplasm transfers in tobacco (Horlow et al., 1993).

However, this mutation does not exist in other species and could be used only in those where grafting is possible. Therefore, it was necessary to develop other markers which could permit the selection of haploids based on procedures which mimic the Rac- allele phenotype.

2.3.3 Conditional dominant lethal markers

A conditional lethal marker leads to the arrest of growth and/or the death of an individual, only in the case where very specific conditions are effective, such as, for example, the addition of a precursor molecule which is converted by this gene into a toxic compound (Fig. 2.2).

It was demonstrated that an overproduction of auxin inhibits root formation of plantlets that express the *tms2* or *aux2* genes when they are growing in a medium supplemented by IAM or NAM (Budar et al., 1988), whereas wild type plantlets are insensitive to the NAM or IAM and grow normally. The *aux2* and *tms2* genes constitute counterselectable markers and were introduced into the genome of crops such as *Brassica,* tobacco and tomato.

Transgenic tomato plants with the *aux2* gene did not show any alteration in their morphology and the formation of the root system was not affected.

Few assays *in vitro* were performed to determine the NAM concentration necessary to inhibit root system formation. Rapidly, some problems of "cross-feeding" have appeared and because of the low frequency of haploid production, several thousands of seeds should be sown. For theses reasons, a procedure of selection was developed in the greenhouse conditions. After determination of the NAM concentration added in the nutrient solution used to water the plants, 7 500 tomato class 2 seeds were sown, 10 plants developed normally on 40mg/l NAM. Three of them exhibited the recessive morphological markers of the maternal parent, and were haploid plants (confirmation by chromosome counting); whilst the 7 other plants displayed the morphological phenotype of the paternal parent, and appeared to be diploid plants which escaped from selection (Hamza et al., 1993).

In parallel to this method of selection, another haploid selection was performed simply by screening the recessive morphological markers of the maternal parent. From 15,000 class 1 seeds, and 1,500 class 2 seeds, 4 and 3 gynogenetic haploids were selected respectively . The rate of haploid plants selected from class 2 seeds was 5-fold lower using the *aux2* gene as counterselectable marker than that obtained by the visual screening of recessive morphological markers of the

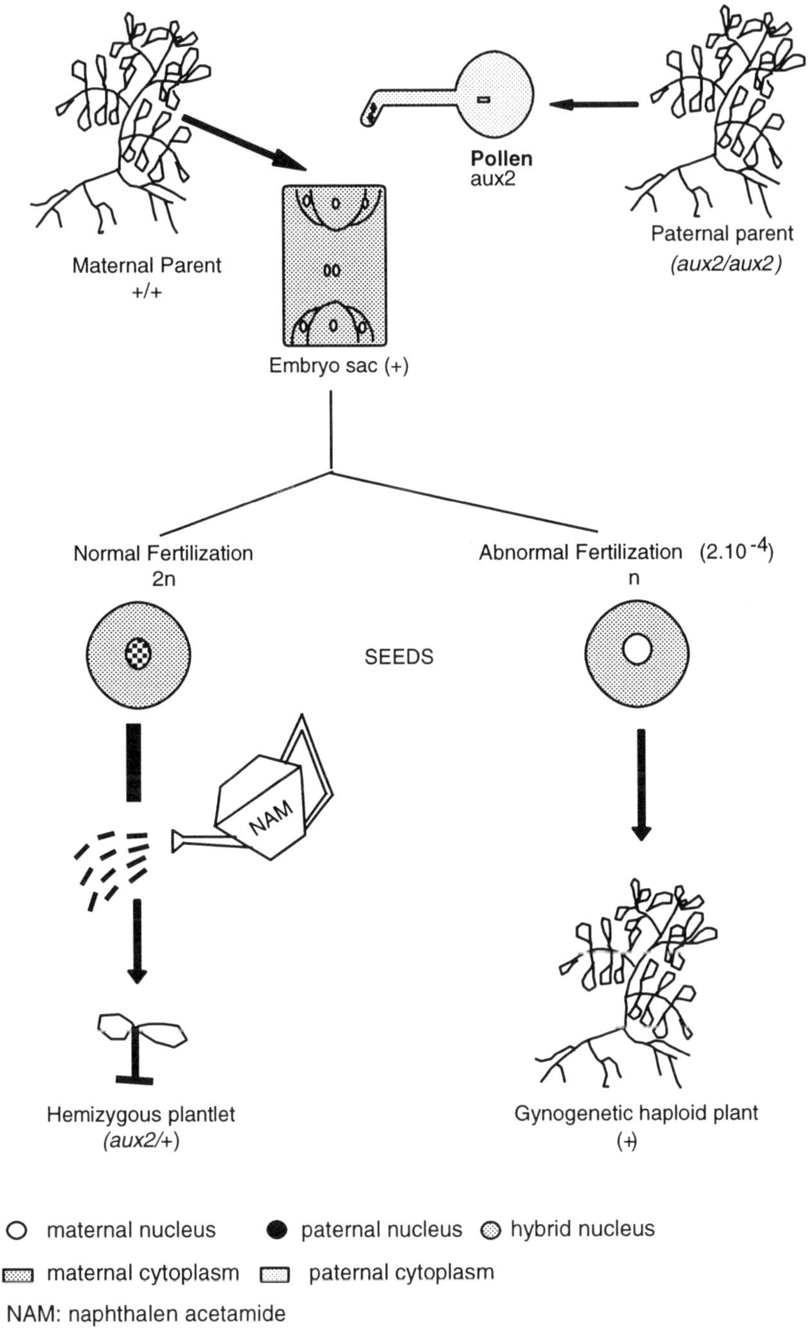

Fig. 2.2. Scheme of spontaneous gynogenetic haploid selection by using aux 2 gene, adapted from the thesis of S. Hamza, 1992.

maternal parent. Thus, although the *aux2* gene seems to be efficient, for reasons of "cross-feeding" between the *aux2* hybrid plants and haploids, it is not the ideal marker for the selection of tomato haploid plants. Haploid plants are less vigourous than diploid and are more sensitive to the toxic effect of NAA. One solution to these problems would be to increase the expression level of the *aux2* gene in transgenic plants which should confer a higher sensitivity to NAM, and should permit selection of haploid plants on lower NAM concentrations.

In *Brassica,* parthenogenetic haploid selection based on the use of aux2 gene has been possible *in vitro* on medium containing NAM. The problem of "cross-feeding" was overcome by increasing the volume of culture media (personal communication, Bechtold, 1995).

Despite some problems, the *aux2* gene is to our knowledge the first counterselectable marker which allowed the selection of tomato haploids (Hamza et al., 1993).

2.4 Conclusions and outlook

The principal feature of current haploidization techniques is the heterogeneity of the results.

Concerning the techniques of haploidization themselves:
Haploids can now be obtained in a large number of species and many possible pathways have been successfully developed. Anther culture and interspecific hybridization are the most efficient and widely used, but they are not universal.

Concerning the responsive species:
At least four techniques can efficiently be used to haploidize barley: androgenesis, interspecific hybridization, spontaneous gynogenesis under genetic control and *in vitro* ovule culture. On the contrary there is still no satisfying method to obtain tomato haploids.

Concerning the potential use of haploids:
The production of haploids without restriction remains limited to a very small number of species, but this condition is probably not essential for their use, especially in the case of nucleo-cytoplasmic exchanges.

Thanks to the already most efficient techniques, especially anther and pollen culture, progress in mass production have been made. More progress is still expected from studies of the genetic control of androgenesis and on the physiology of the first steps of pollen embryogenesis. Though these two ways appear in competition, they are in fact complementary. A better knowledge of the conditions of embryogenesis may increase the number of responsive genotypes. Reciprocally, a selection for responding genotypes can boost the proportion of embryogenic microspores.

Beside these techniques, some other pathways still could be investigated:
The chemical treatments of pollen before fertilization have produced poor results to date, but the inhibition of the second division of the generative nucleus by antimitotics could constitute a "pseudo phureja" method for non responsive tetraploid species.

The fusion of pollen protoplasts, which are unable to divide, with inactive somatic protoplasts could be another possibility for producing homozygous plants with nucleo-cytoplasmic exchanges, (Desprez et al., 1995).

Acknowledgements

The authors appreciate having this opportunity to honor Jean-Pierre Bourgin for his pioneering work on haploidy. We thank Jacqui Shykoff for helpful comments and manuscript corrections.

References

Ao GM, Zhao SX, Li GH, (1982) In vitro induction of haploid plantlets from unpollinated ovaries of corn, Zea mais L. Acta Gen Sin 9:281 – 283

Asselin de Beauville M (1980) Obtention d'haploides in vitro à partir d'ovaires non fécondés de Riz, Oryza sativa L. CR Acad. Sci. Paris 296:489 – 492

Baenziger PS, Kudirka DT, Schaeffer GW, Laza MD (1984) The significance of doubled haploid variation. In: JP Gustafson (ed.), Gene Manipulation in Plant Improvement, Plenum Press, New York, pp 385 – 414

Bajaj YPS (1990) In vitro production of haploids and their use in cell genetics and plant breeding. In: Bajaj YPS (ed) Biotechnology in agriculture and forestry, vol 12, Haploids in Crop Improvement 1. Springer, Berlin, Heidelberg, New York, pp 3 – 11

Bennet MD, Finch RA, Barclay IR (1976) The time rate and mechanism of chromosome elimination in Hordeum hybrids. Chromosoma 54:175 – 200

Bentolila S, Hardy T, Guitton C, Freyssinet G (1992). Comparative genetic analyses of F2 plants and anther culture derived plants of maize. Genome 35:575 – 582

Blakeslee A, Belling J (1924) Chromosomal mutations in the Jimson weed, Datura stramonium. J. Hered 15:195 – 206

Blakeslee A, Belling J, Farnham ME, Berger AD (1922) A haploid mutant in the Jimson weed Datura stramonium. Science 55:646 – 647

Bourgin JP (1978) Valine-resistant plants from in vitro selected tobacco cells. Mol Gen Genet 161:225 – 230

Bouvier L, Zhang YX, Lespinasse Y (1993) Two methods of haploidization in pear (Pyrus communis L.): greenhouse seedling selection and in situ parthenogenesis induced by irradiated àollen. Theor Appl Genet 87:229 – 232

Bozogipour R, Snape KV (1992) The relationship between in vitro performance of haploid embryo and the agronomic performance of the derived doubled haploid line in Barley. Theor Appl Genet 84:118 – 122

Budar F, Deboeck F, Van Montagu M, Hernalsteens JP (1988) Introduction and expression of the octopine T-DNA oncogenes in tobacco plants and their progeny. Plant Sci 46:195 – 206

Burk LG, Gerstel DU, Wernsman EA (1979) Maternal haploid of Nicotiana tabacum L. from seed. Science:206, 585

Burk LG (1962) Haploids in genitically marked progenies of tobacco. J Hered 53:222 – 225.

Cai DT, Zhou C, (1984). In vitro induction of haploid embryoids and plantlets from unpollinated young florets and ovules of Helianthus annuus L. Kexue Tonbao 29, pp 680 – 682

Campion B, Azzimonti MT (1988) Evolution of ploidy level in haploid plants of onion Allium cepa obtained through in vitro gynogenesis. In: 4th EUCARPIA Allium Symposium, Wellesbourne (UK), 6 – 9 September 198 pp 85 – 89

Chase SS (1969) Monoploids and monoploids-derivatives of maize (Zea mays L.). Bot Rev 35:117 – 167

Chase SS (1963) Androgenesis – its use for transfer of maize cytoplasm. J Hered 54:152 – 158

Chase SS (1952) Production of homozygous diploids of maize from monoploids. Agron J,44:263 – 267

Chase SS (1951) The monoploid method of developing inbred lines. Proc 6th Annual Hybrid Corn Industry – Research Conf, pp 29-34

Chen BY, Heenen WK (1989) Evidence for spontaneous diploid androgenesis in Brassica napus L. Sex. Plant Reprod 2:15-17

Chen ZH, Li WB, Zhang LH, Xu XE, Zhang SJ (1988) Production of haploid plantlets in cultures of unpollinated ovules of Hevea brasiliensis Muell.-Arg. In: M.R. Ahuja (ed.) Proc IUFRO Working Party S2.04 – 07, Grosshansdorf, FRG, 10 – 13 August 1987. Kluwer Academic Publishers, Dordrecht, pp 39 – 44

Clausen RE, Lammerts WE (1929) Interspecific hybridization in Nicotiana. Haploid and diploid merogony. Am Nta 43:279 – 282

Coe EH, Sarkar S (1964) The detection of haploids in maize. J Hered 55:231 – 233

Coe EH, (1959) A line of maize with high haploid frequency. Am Nat 93:381 – 382

Cohat J (1994) Obtention chez l'échalotte (Allium cepa L. var aggregatum) de plantes haploïdes gynogénétiques par culture in vitro de boutons floraux.. Agronomie 14:299 – 304

Deaton WR, Keyes GJ, Collins GB (1982) Expressed resistance to black shank among tobacco callus culture. Theor Appl Genet 63:65 – 70

De Buyser J, Henry Y, Lonnet P, Herzog R, Hespel A (1987) “Florin”: a doubled haploid wheat variety developed by the anther culture method. Plant breeding 98:53 – 56

De Verna JW, Collins GB (1984) Maternal haploids of Petunia axillaris Lam. via culture of placenta attached ovules. Theor Appl Genet 69:187 – 192

Desperez B, Chupeau MC, Vermeulen A, Delbreil B, Chupeau Y, Bougin JP (1995) Regeneration and characterisation of plants produced from mature tobacco pollen protoplasts via gametosomatic hybridization. Plant Cell Reports 14:204 – 209

Dore C (1989) Obtention de plantes haploides de chou cabus (Brassica oleracea L. ssp. capitata) après culture in vitro d’ovules pollinisés par du pollen irradié. CR Acad Sci Paris 309D:875 – 878

Dore C, Marie F (1993) Production of gynogenetic plants of onion (Allium cepa L.) after crossing with irradiated pollen. Plant Breed 111:142 – 147

Dumas de Vaulx R (1979) Obtention de plantes haploïdes chez le melon (Cucumis melo L.) après pollinisation par Cucumis ficifolius A. Rich. C Acad SC. Paris, 289, Série III, pp 875 – 878

Goodsell SF (1961) Male sterility in corn by androgenesis. Crop Sci, 1:227 – 228

Gu ZP, Cheng KC (1983) In vitro induction of haploid plantlets from unpollinated young ovaries of lily and embryological observations. Acta Bot. Sin 25:24 – 28

Guha S, Maheswari SC (1964) In vitro production of embryos from anthers of Datura. Nature, 204:497

Guha S, Maheswari SC (1966) Cell division and differenciation of embryos in the pollen grains of Datura. Nature 212:97 – 98

Hamza S, Camilleri C, Pollien JM, Vaucheret H, Bourgin JP, Chupeau Y (1993) Selection for spontaneous tomato haploids using a conditional lethal marker. Theor Appl Genet 86:657 – 664

Heberle-Bors E, Reinert J (1979) Androgenesis in isolated pollen cultures of Nicotiana tabacum. Dependance upon pollen development. Protoplasma 99:237 – 245

Henry Y, Vain P, De Buyser J (1994) Genetic analysis of in vitro plant tissue culture responses and regeneration capacities. Euphytica 79:45 – 58

Hoekstra S, Van Zijderwald MH, Louwerse JD, Heidekamp F, Van der Mark F (1992) Anther and microspore culture of Hordeum vulgare L. cv. Igri. Plant Sci 86:89 – 96

Horlow C, Defrance MC, Pollien JM, Goujaud J, Delon R, Pelletier G (1993) Transfer of cytoplasmic male sterility by spontaneous androgenesis in tobacco (Nicotiana tabacum L.). Euphytica 66:45 – 53

Hosemans D, Bossoutrot D (1983) Induction of haploid plants from in vitro culture of unpollinated beet ovules (Beta vulgaris). Z Pflanzenzüchtg 91:74 – 77

Hougas R W, Peloquin SJ, Gabert AC (1964) Effect of seed parent and pollinator on frequency of haploids in Solanum tuberosum. Crop Sci 4:593 – 595

Illies ZM (1974) Induction of haploid parthenogenesis in aspen by post-pollination treatment with toluidine blue. Sylvae Geneticae 23:221 – 226

Illies ZM (1964) The occurence of haploid seedlings in P. abies. Naturwissenschaften 51:442

Kappert H (1950) Botanische untersuchiengen zur Erblichkeit der polyembryonie. Moderne Biologie. FW Peters, Berlin

Kappert H (1933) Erbliche polyembryonie bei Linum usitatissimum. Biol Zentralblatt 53:276 – 307

Kasha KJ, Kao KN (1970) High frequency haploid production in barley (Hordeum vulgare L.). Nature 225:874 – 876

Kaul MLH (1986) Male sterility in doubled haploid rice. Rice Genet Newsl, 3:109

Keller ERJ, Korzun L (1996) Ovary and ovule culture for haploid production. In Mohan Jain S, Sopory SK, Veilleux RE, (eds) In vitro Haploid Production in Higher Plants, vol 1, Fundamental Aspects and Methods. Kluwer academic publishers, Dordrecht, Boston, London, pp 217 – 235

Kermicle JL (1969) Androgenesis conditioned by a mutation in maize. Science 166:1422 – 1424

Kostoff D (1929) An androgenetic Nicotiana haploid. Zeits. Zellforsch 9:640 – 642

Lacadena JR (1974) Spontaneous and induced parthenogenesis and androgenesis. In: Guelph, Kasha (ed) Haploids in higher plants. Advances and potential. Proceed of the 1st International Symposium p 13 – 32

La Rue CD (1948) Regeneration in the megagametophyte of Zamia floridana. Bull Torrey Bot Club 75:597 – 603

Laurie DA, Bennett MD (1988) The production of haploid wheat plants from wheat x maize crosses. Theor Appl Genet 76:393 – 397

Leviel C, Huyghe C (1985) Observation des gamètogenèses mâle et femelle, de la fécondation et de la formation d'embryons non-zygotiques après éclaircissage des anthères et sacs embryonnaires de Cichorium intybus L. et de Linum usitissimum L. C R Acad Sci Paris 20:769 – 774

Lichter R (1982) Induction of haploid plants from isolated pollen of Brassica napus. Z. Pflanzenphysiol 105:427 – 434

Mascarenhas JP (1989) The male gametophyte of flowering plants. Plant Cell, 1:657 – 664

Maheswari SC, Tyagi AK, Malhotra K, Sopory SK (1980). Induction of haploidy from pollen grains in angiosperms - the current status. Theor Appl Genet 58:193 – 206

Meynet J, Barrade R, Duclos A, Siadous R (1994) Dihaploid plants of roses (Rosa x hybrida, cv. "Sonia") obtained by parthenogenesis induced using irradiated pollen and in vitro culture of immature seeds. Agronomie 14:169 – 175

Mohan Jain S, Sopory SK, Veilleux RE (1996) In: In vitro Haploid Production in Higher Plants, Vol. 1, Fundamental Aspects and Methods. Kluwer academic publishers, Dordrecht, Boston, London, 356p

Murigneux A, Barloy D, Leroy P, Beckert M (1993) Molecular and morphological evaluation of doubled haploid lines in maize. 1. Homogeneity within DH lines. Theor Appl Genet. 86:837 – 842

Niemirowicz - Scytt K, Dumas de Vaulx R (1989) Preliminary data on haploid cucumber (Cucumis sativus L.) induction. Cucurbitae Genetics Coop 12:24

Nitsch C, Norreel B (1974) La culture du pollen isolé sur milieu synthétique. CR Acad Sci Paris 278:1031 – 1034

Nitsch C, Norreel B (1973) Effet d'un choc thermique sur le pouvoir embryogène du pollen de Datura innoxia cultivé dans l'anthère et isolé de l'anthère. C R Acad Sci Paris 275:303 – 306

Pandey KK, rzywara L, Sanders PM (1990) Induced parthenogenesis in kiwifruit (Actinidia deliciosa) through the use of lethally irradiated pollen. Euphytica 51:1 – 9

Pelletier G, Ferault M, Goujaud J, Vedel F, Caboche M (1987) The use of rootless mutants for the screening of spontaneous androgenetic and gynogenetic haploids in Nicotiana tabacum: evidence for the direct transfert of cytoplasm. Theor Appl Genet 75:13 – 15

Powell W, Caligari PDS, Hayter AM (1983) The use of pollen irradiation in barley breeding. Theor Appl Genet 65:73 – 76

Rao PS, Suprasanna P (1996) Methods to double haploid chromosome numbers. In:Mohan Jain S, Sopory SK, Veilleux RE (eds.) In vitro Haploid Production in Higher Plants, vol 1, Fundamental Aspects and Methods. Kluwer academic publishers, Dordrecht, Boston, London:317 – 339

Raquin C (1985) Induction of haploid plants by in vitro culture of Petunia ovaries pollinated with irradiated pollen. Z. Pflanzenzüchtg 94:166 – 169

Raquin C, Cornu A, Farcy E, Maizonnier D, Pelletier G, Vedel F (1989) Nucleus substitution between Petunia species using gamma ray induced androgenesis. Theor Appl Genet 78:337 – 341

Rode JC, Dumas de Vaulx R (1987) Obtention de plantes haploïdes de carottes (Daucus carota L.) issues de parthénogenèse induite par du pollen irradié et culture in vitro des graines immatures. CR Acad Sci Paris 305D: 225 – 229

San Noeum LH (1976) Haploides d'Hordeum vulgare L. par culture in vitro d'ovaires non fécondés. Ann Amélior Plantes 26:751 – 754

San LH, Demarly Y (1984) Gynogenesis in vitro and biometric studies of doubled haploids obtained by three techniques in Hordeum vulgare L. In:Lange W, Zeven AC, Hogenboom NG (eds) Efficiency in Plant Breeding. Proc 10th Congr EUCARPIA Wageningen, 19 – 24 June 1983, p 347

Sangwan RS, Sangwan BS (1986) Effets des rayons gamma sur l'embryogénèse somatique et l'androgénèse chez divers tissus végétaux cultivés in vitro. In: Nuclear techniques and in vitro culture for plant improvement. IAEA, Vienna, pp 181-185.

Sarkar KR, Sachan JKS, Mathur DS (1972) Detection of androgenetic monoploids in maize. Curr Sci 41:190 – 192

Sauton A, Dumas de Vaulx R (1987) Obtention de plantes haploï des chez le melon (Cucumis melo L.) par gynogenèse induite par du pollen irradié. Agronomie 7:141-148.

Sitbon M (1981) Production of haploid Gerbera jamesonii plants by in vitro culture of unfertilized ovules. Agronomie 1:807 – 812

Snape JW, Parker BB, Simpson E, Ainsworth CC, Payne PI, Lawc N (1983) The use of irradiated pollen for differential gene transfer in wheat (Triticum aestivum). Theor Appl Genet 65:103 – 111

Straub J (1958) Das Überwinden der Selbststerilität. Z Bot 46:98 – 111

Subrahmanyam NC, Kasha KJ (1973) Selective chromosomal elimination during haploid formation in barley follwing interspecific hybridization. Chromosoma 42:111 - 125

Sunderland M, Xu ZH (1982) Shed pollen culture in Hordeum vulgare. J Exp Bot, 33:1086 – 1095

Symko S (1969) Haploid barley from crosses of Hordeum bulbosum (2X) x Hordeum vulgare (2X). Can J Genet Cytol 11:602 – 608

Tanskley SD, Zamir D, Rick CM (1981) Evidence extensive overlap of sporophytic and gametophytic gene expression in Lycopersicon esculentum. Science 213:453 – 455

Tao ZR, Liu MS, Zhu SC (1985) In vitro production of haploid plantlets from unpollinated ovules of potato. Hereditas 7:5

Thevenin L (1968) Les problèmes d'amélioration chez Asparagus officinalis. II. Haploïdie et amélioration. Ann Amel Plantes 18:327 – 365

Truong-Andre I (1988) In vitro haploid plants derived from pollination by irradiated pollen of Cucumber. In: INRA (ed) Proceeding Eucarpia Meeting on Cucurbit Genetic and Breeding, Avignon Montfavet, France, pp 143 – 144

Wenzel G, Hoffmann F, Thomas E (1976) Heterozygous microspore derived plants in Rye. Theor Appl Genet 59:333 – 340

Wu BJ, Chen KC (1982) Cytological and embryological studies on haploid plant production from cultured unpollinated ovaries of Nicotiana tabacum. Acta Bot Sin 2:125 – 129

Wu KX, Xu MZ (1984) Induction of maternal haploid from unpollinated ovaries of poplar in vitro. Acta Gen Sinica 11:47 – 51

Willing RP, Mascarenhas JP (1984) Analysis of the complexity and diversity of an RNA from pollen and shoots of Tradescantia. Plant Physiol 75:865 – 868

Ye JM, Kao KN, Harvey BL, Rossnagel BG (1987) Screening salt-tolerant barley genotypes via F1 anther culture in salt stress media. Theor Appl Genet 74:426 – 429

Zhang YX, Lespinasse Y, Chevreau E (1988) Obtention de plantes haploïdes de pommier (Malus x domestica Borkh.) issues de parthénogenèse induite in situ par du pollen irradié et culture in vitro des pépins immatures. CR Acad Sci Paris 307 série III 451-457

Zhu ZC, Wu HS (1979) In vitro induction of haploid plantlets from the unpollinated ovaries of Triticum aestivum and Nicotiana tabacum. Acta Gen Sinica 6:181-183

3 Genetic Analysis of in vitro Androgenetic Response in Maize

M. Beckert
Station d'Amélioraton des Plantes, INRA, 234, Avenue du Brézet,
Domaine de Crouelle, 63039 Clermont Fd Cedex, France

3.1 Introduction

The potential benefits of tissue culture techniques in plant breeding have often been limited by the difficulty in establishing *in vitro* cultures that present efficient plant regeneration capacity. Part of these difficulties are connected to the differential genotypic response. Genotypic variation for general tissue culture ability is evident in the major cereal crops and these aspects have been widely described during the past years and reported as in wheat (Sears and Deckard 1982), corn (Beckert and Pollacsek 1979, Duncan et al. 1985, Hodges et al 1986), rice (Guha-Mukherjee 1973, Reddy et al. 1985), barley (Hanzel et al. 1985), triticale (Nakamura and Keller 1982), sorghum (Ma et al. 1987) and oats (Cummings et al. 1976). These publications cover morphogenesis as well as somatic embryogenesis, but also other *in vitro* traits.

The genetic component of anther culture response has received less attention despite being one of the most important factors limiting the widescale application in breeding programs for dihaploid plant production. Genetic analysis of anther culture response has also been carried out in major cereals crops such as wheat (Agache et al.1989, Becraft and Taylor 1992), barley (Foroughi-Wehr et al 1982), rice (Miah et al. 1985), maize (Petolino and Thompson 1987). Formely, quantitative genetic analyses using, for instance, diallel trials, were conducted to determine the inheritance of *in vitro* traits. Recently, using recombinant monosomic wheat lines, genetic components of some tissue culture traits have been allocated to some specific chromosomic wheat or rye segments. Now, the use of precise molecular genetic maps allows us to identify more precisely some major QTL involved in regeneration traits.

The following discussion adresses the current status of anther culture in maize. Emphasis is placed on the genetic analysis of the different *in vitro* characters leading to the initiation of multicellular structures from the microspores, androgenetic embryo formation, their regeneration as plantlets and spontaneous plant chromosome doubling. A possible application of anther culture technology in classical maize breeding will also be discussed.

3.2 Basic maize anther culture techniques

Since the pioneering work of Chinese scientists, who were the first to obtain androgenetic maize plants through anther culture (Anonymous 1975), numerous other teams (Miao et al. 1981; Genovesi and Collins 1982; Dieu and Beckert 1986; Petolino and Jones 1986) have worked on the characterization of tissue culture conditions, including optimal growth conditions for donor plants, anther pretreatments, media... leading to a higher global androgenetic response. Briefly, we summarize here the main technical aspects of the process.

Tassels with anthers containing microspores at the uninucelate/late uninucleate stages are the most responsive. Precise cytological observations and/or size and color of the anthers lead to a correct identification of the anthers to be plated. If necessary, a staging procedure is carried out using Alexander staining (Alexander, 1969). Cold pretreatment of the microspores, stimulating androgenesis, is applied directly on the anthers, after their rapid plating for about two to three weeks in the dark at 8°C. Petri dishes are then transferred to the standard condition at 28°C under low light intensity. After three weeks, the anthers are manually transferred onto a new medium, with a lowered sucrose concentration, from 120g/l to 25g/l, and modified hormonal composition from TIBA (0.1mg/l) to Kinetin (1mg/l), to promote a direct regeneration or germination process of the androgenetic embryos. The precise media compositions are those described in Barloy and Beckert (1993). Regenerated small plantlets are transferred to a simple medium for rooting and then to pots. Only plantlets recovering the normal set of chromosomes, through spontaneous chromosome doubling, are transferred to the greenhouse after cytometric cellular DNA check. A general summary of the whole process and timing is given at the Fig 3.1.

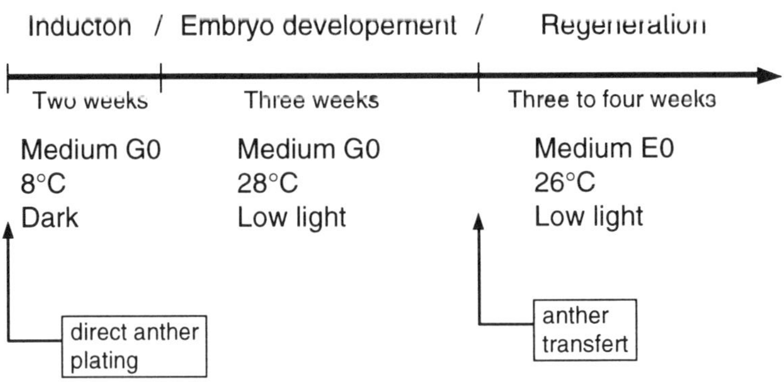

Fig. 3.1. Timing of maize *in vitro* androgenesis)

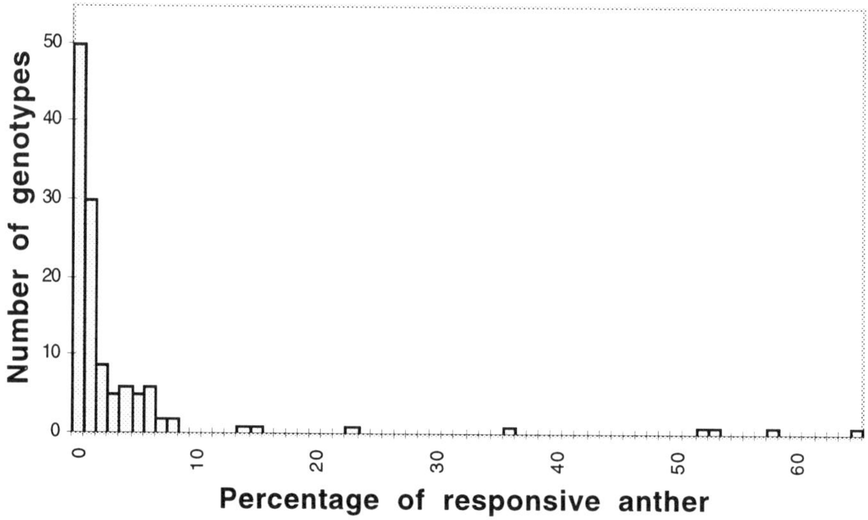

Fig. 3.2. Expression of the genetic variability for anther culture response in maize

3.3 Genetic variability for anther culture response within maize

3.3.1 Percentage of anther culture response

Maize is characterized by a very large genetic variability which has been described for many different agronomic and/or morphological characters. Here, anthers of various genotypes have been plated following the protocole described in the Figure 3.1. The genotypes were chosen following their different genetic or geographic origins. Fig. 3.2 represents a histogram of the response of these genotypes for the percentage of their anthers giving at least one macroscopic androgenetic embryo.

Although a large part of the genetic variability of maize appears to present no, or only a very low aptitude to form androgenetic embryos (frequence of responding anthers lower than 1 to 2 percent), a few origins present a very high and interesting level of reactivity (more than 50% of responding anthers).

3.3.2 Other anther culture reactivity characters

Table 3.1 summarizes the main characters to be taken into account when plating anthers from a practical point of view. Although the genotypes reported here were selected from previous anther culture experiments, they include also elite lines selected for their agronomic value. The synthetic population SDB developed in our plant breeding station contains a wide range of different inbred lines, such as the lines Pa91, H99 and Fr16, previously described by Petolino and Jones (1986), and other agronomically elite lines. The single cross hybrid A16xQ3, related to A188, B73 and DH5 was selected for its very high embryogenic somatic tissue culture response. The lines B14 and B64 are related to the B14 group, Oh43 to the Lancaster, Nay 565 and R158 to the Illinois High Yield group, and Pb116 was selected from flint material in Portugal. The pure lines F1937, F1954 and DH147 are related to the Iodent group. The dihaploid lines DH5 and DH7 are of Chinese origin.

Table 3.1. Global characterization of maize genetic variability for *in vitro* androgenetic response.

Variable Genotype	AC	ELS	PLA	F2n	DH%
SDB	7.79	30.89	2.37	42.18	1.00
A16xQ3	10.50	17.12	5.54	37.04	2.05
B14xOh43	8.53	38.96	6.09	23.19	1.41
B64xPb116	6.04	18.49	4.60	13.33	0.61
NaY565xR158	5.66	11.15	2.51	30.61	0.80
F1937xDH229	38.06	207.89	34.04	21.52	7.32
F1954xDH99	30.20	210.90	12.91	16.15	2.08
DH147xL1	35.18	133.46	17.62	19.23	3.39
A188xDH7	32.95	109.52	32.95	27.35	4.60

AC, Percentage of anther giving at least one macroscopic androgenetic embryo;
ELS, Number of macroscopic embryo like structure produced for one hundred plated anthers;
PLA, Number of regenerated plantlets for one hundred plated anthers;
F2n, Frequence of spontaneous dihaploid plantlets;
DH%, Number of dihaploid plants for one hundred plated anthers.

On this restricted sample of selected genotypes, it is clear that there is also genetic variability for other *in vitro* characters such as number of embryo-like structures developed for 100 plated anthers (ELS) which vary from 10 to more than 200 in the most favorable case. In the same manner, the number of plantlets regenerated for 100 plated anthers (PLA) and the frequency of spontaneous chromosome doubling of regenerated plantlets (F2n) seem to be affected as much as the number of dihaploid plantlets regenerated for 100 plated anthers (DH%).

3.4 Genetic analysis

3.4.1 Lines versus F1 observation

Evaluation of anther culture responsiveness of inbred lines is presented in Table 3.2. These results were obtained with a large number of anthers plated from different tassels, representing replications of the same genotype. The evaluation of the behaviours of the DH lines (DH5, DH7, DH9), derived through androgenesis from three different unrelated genetic backgrounds (Barloy et al. 1989) reveals a very large variation in response. Also, the fact that the DH9 line presents a null response for androgenesis, is in accordance with the hypothesis that the anther wall of the donor plant is at least partially involved in the androgenetic induction pathway. The agronomic inbred lines evaluated (F244, Co220, F1656) do not reveal any anther culture response.

Table 3.2. General comparison of lines and hybrid counterpart values.

Variable Genotype	AC	ELS	PLA
DH5	31.2	293.0	5.2
DH7	28.5	364.0	11.7
DH9	0.0	0.0	0.0
F244	0.1	0.5	0.0
MBS847	0.0	0.0	0.0
Co220	0.0	0.0	0.0
F1656	0.0	0.0	0.0
DH5xDH7	66.2	458.0	13.0
F244xDH5	15.4	52.2	1.5
F244xDH7	19.0	92.4	3.6
MBS847xDH5	9.1	52.9	1.1
MBS847xDH7	5.5	12.5	0.0
Co220xDH5	13.3	25.8	0.9
Co220xDH7	14.2	26.7	1.1
F1656xDH5	0.5	0.5	0.0
F1656xDH7	3.0	5.0	0.0
BMS	0.0	0.0	0.0
ND203	0.0	0.0	0.0
BMSxND203	17.0	30.3	1.1

The hybrid DH5xDH7 reveals a considerable heterosis in comparison with the parental value, leading to the idea of separate genetic factors in the two lines showing good complementation. Although the different agronomic elite lines evaluated do not reveal any aptitude of this level, their hybrid counterparts with a high DH responding line present an intermediate value close to the mean value of both parents. Only the line F1656 seems to strongly depress the anther reactivity of the hybrid. Table 3.2 also reports, in the last three lines, a clear case of genetic complementation between the genetically unrelated lines BMS and ND203.

The main conclusion derived from this study is that the genetic control of androgenetic response is certainly not recessive, but that the general genetic context of the parents may modulate its expression in hybrid structures.

3.4.2 F2 versus DH lines observations

This first step in genetic analysis was continued with a more precise study including an evaluation of a set of recombinant DH lines derived from the hybrid DH5xDH7 using androgenesis. Another sample of recombinant DH lines derived from the hybrid A188xDH7, developed between a recalcitrant and a highly responding line, was also evaluated using the same general protocole.

All the recombinant DH lines derived from these two genetic bases were investigated for their androgenetic reactivity including also the regeneration steps. The main advantages in this strategy, compared to a classical F2 segregation analysis, lies in the fact that it was possible, for a particular DH line, to repeat the same biological test on different individual plants. This leads to a better and more precise evaluation of its genetic capacity. The global results concerning the percentage of responsive anthers are presented in Figures 3.3a and 3.3b.

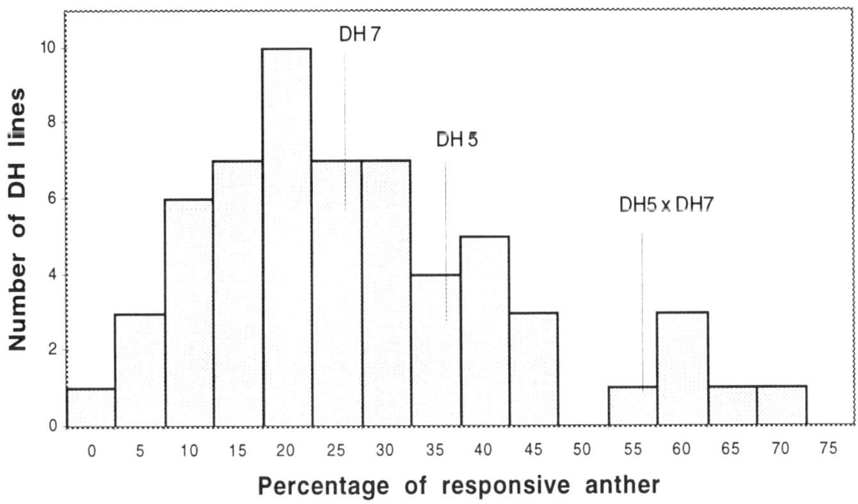

Fig. 3.3a. Genetic segregation for androgenetic capacity in DH5 x DH7 hybrid

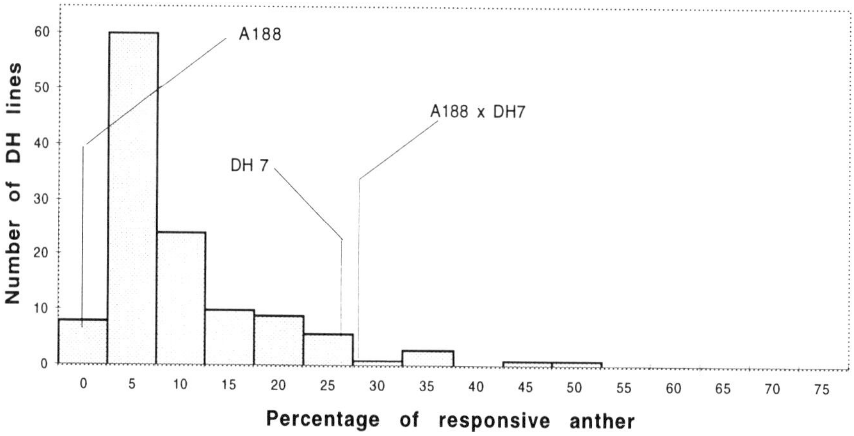

Fig. 3.3b. Genetic segregation for androgenetic capacity in A188 x DH7 hybrid

A very large variation, from 0 to 70% of responding anthers, is observed between the DH lines for both of these crosses. From this data, it appears that the androgenetic process does not necessarily select for higher response level, since from DH5xDH7, DH lines with the high responsiveness of the hybrid were obtained as well as lines with very little or null potential. The same observation is recorded for the second genetic origin.

Although the line A188 does not respond to androgenesis and DH7 presents only about 30% of responsive anthers, it was possible to obtain some transgressive lines giving rise to 50% of responding anthers. This strongly suggests that A188, the nul responding line, has some positive alleles that are complementary to those of DH7.

Although DH5 and DH7 are genetically unrelated lines, it was possible to fix the high responsiveness of the hybrid in a limited number of recombinant lines. Therefore, the distributions and ranges of variation of these samples indicate that a small number of genes, or group of genes may lead to the differences between parent lines in both crosses. Using Snape's formula and according to restrictive hypothesis (Snape et al. 1984), it appears that only four to five groups of genes, different between parents, may explain the variation observed.

The second step of the androgenetic pathway concerns androgenetic embryo capability to regenerate as plantlets. Figure 3.4 represents part of the genetic relations between two characters (percentage of responding anthers and capability of embryos to regenerate) for the DH5 x DH7 recombinant DH lines.

Within this cross, in this set of the 2nd cycle of DH androgenetic lines, it was possible to recombine a high level of androgenetic anther response with a high capability of plant regeneration. This last character was improved through tissue culture step due to the observation of transgressive lines in comparison with both values of DH5 and DH7 lines and the value of the hybrid. Implementing the same Snape's

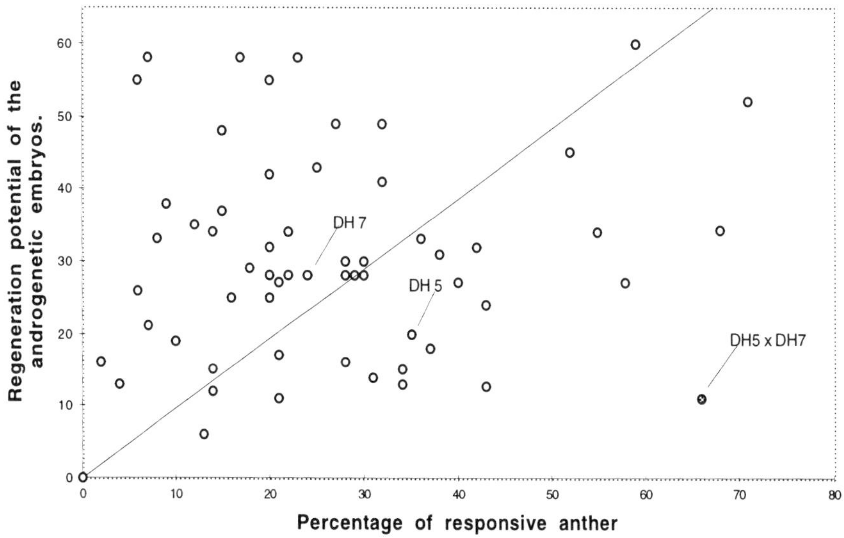

Fig. 3.4. Relationship between percentage of responsive anther and regeneration potential

formula as previously used reveals also that this regeneration character is probably determined by a limited number of genes differing between parents.

3.4.3 Segregation distortion

In androgenetic populations of various crop plants, several authors have reported segregation distortions for a range of genetic markers (mophological, biochemical, molecular...). This has been observed for barley (Thompson et al. 1991), rice (Guiderdoni 1991), and wheat (Henry et al. 1993). The development of molecular markers such as RFLP or PCR based markers offers the opportunity to amplify these investigations. The establishment of genetic maps offers also the opportunity for investigations on the whole genome level. With maize, the comparison of sample of DH lines derived from crosses involving either one or two anther culture responsive parental lines had been made. This, made possible the evaluation of the consequences of different responsiveness on segregation ratios.

Two populations of 80 DH (DH5xDH7 and A188xDH7) and one of 80 SSD F6 lines (A188xDH7) were tested using more than 100 spaced RFLP probes (Murigneux et al., 1993). Distortions in segregation ratios were found in only seven loci over 99 tested for the SSD lines, with half towards each parent for the A188xDH7 cross. In contrast, the distortions found in DH lines were more abundant and clearly different between the two DH populations. Deviations from the 1:1 ratio were much larger in the DH population where only one parent was responsive to androgenesis. The distortions were less frequent in the DH5xDH7 population where the two parents were highly androgenetic lines. Fig. 3.5, represents a general graphical comparison made between the three populations mentioned. This figure represents, as box plots, the distributions, their medians and the confidence limits of them, for the genome ratio evaluated for the SSD and the DH lines populations. The distribution is skewed towards a high percentage of DH7 allele (androgenetic positive parent) in the A188xDH7 cross for the corresponding DH population. This observation was not recorded in the two other populations where the both parents were responsive, or for the SSD population which was selfed.

Although the main result of this study shows that segregation distortion occurs at high level when considering Mendelian markers, we do not consider this phenomenon as being a real problem in breeding. In fact, the frequency of extreme DH lines statistically close to either one or to the other parent is quite low.

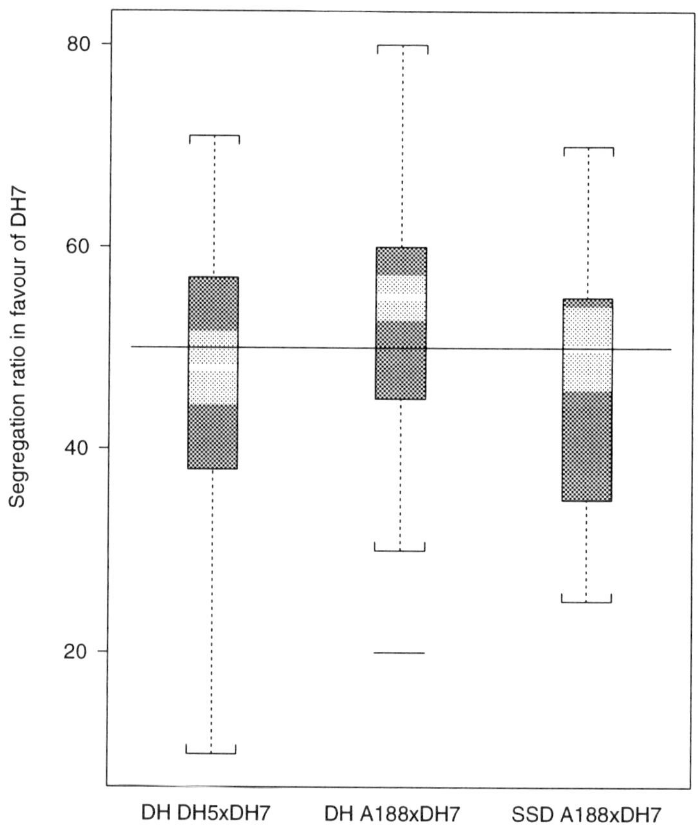

Fig 3.5. Distribution of genomic ratio of sample of lines derived through *in vitro* androgenesis or selfing

3.4.4 QTL for androgenesis

Genetic maps of RFLP markers, closely spaced and covering large parts or even the entire genome, are available for maize. Using different genetic materials and different mating reproduction systems, numerous authors: Armstrong et al. (1992), Cowen et al. (1992), Wan et al. (1992), Murigneux et al. (1994), Beaumont et al. (1995) have been involved in the identification and quantification of QTL effects involved in the androgenetic regeneration capacity of maize.

Considering the A188xDH7 cross, two and three QTL were detected for the percentage of responsive anther and embryo production respectively (see Table 3.3). They explain individually between 10 to 20% of the phenotypic variation. A general statistical model involving all the probes significantly correlated with the anther culture response character account for 40% of the total phenotypic variation. Within this cross, considering the androgenetic embryo production, the same QTL are mapped on the same chromosomes as previously for anther response. This may be interpreted as a consequence of the general correlation between the two *in vitro* traits. Nevertheless, chromosome 1 (probe Bnl 7.08) seems to be only involved for the second step.

Table 3.3. Map position of the QTL involved in maize androgenesis for the A188 x DH7 cross

Character	Probe	Chromosome	LOD/R2	Positive parent
AC	**Adh2 +4**	**4**	**3.17/16.0**	**DH7**
	UMC 166A	**5**	**2.65/13.1**	**A188**
ELS	**Adh2+4**	**4**	**2.34/11.9**	**DH7**
	UMC166A	**5**	**3.93/18.5**	**A188**
	BNL 7.08	**1**	**2.75/13.8**	**A188**

The other main information from this description is the identification of the parent carrying the positive allele. It is clear that the line A188, a recalcitrant line for androgenesis, has positive alleles for both studied characters.

Considering now all the genotypes and different publications together, it is clear that more than twelve chromosomic locations appear to be involved in the androgenetic pathway. Fig. 3.6 summarizes, on a consensus maize map, all the QTL which have been identified. Only chromosome six has never been reported to be involved. From these general observations there are clearly some common chromosomic regions carrying "hot spots" of regeneration, like on the chromosome 8. (The confident limits of the QTL detected in the different works may suggest the same localizations.)

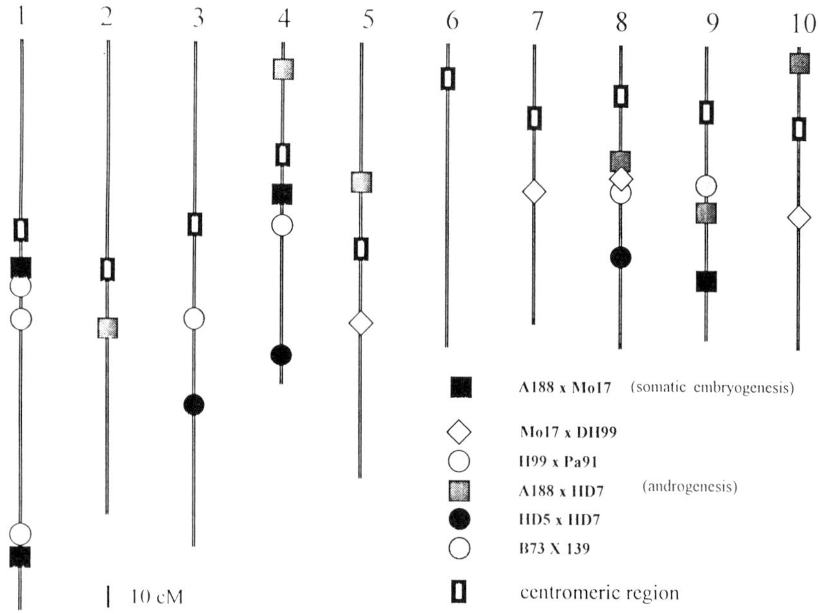

Fig. 3.6. Genotypic variation of QTL involved in in vitro maize regeneration)

3.5 Breeding androgenetic response

To overcome the genetic limitations of *in vitro* androgenesis response it has been proposed to breed for these traits mainly for poor responding material. In particular, Petolino et al. (1988) and Barloy et al. (1990) pointed out that the DH plant production itself may significantly enhance the level of response to anther culture of the derived material.

Using a synthetic population developed at our plant breeding station (cycle C0), anther were plated from a very large number of different individual plants. A set of 30 DH derived from 30 different individual donor plants were intermated to form the next population (cycle C1) and the experiment was repeated to form the final population (cycle C2). Genetic progress for the *in vitro* traits was evaluated through this kind of recurrent selection scheme.The results are reported in Table 3.4.

This table reports, for the different traits evaluated, that the *in vitro* selection had a very strong effect. The mean percentages of responsive anthers, the number of embryo like structures developed or the number of regenerated plantlets produced are increased by a factor 1.4, 2.8 and 3.5, respectively. The percentage of spontaneous chromosome doubling also increased but to a lesser extent than the other characters. This significant increase of 35% confirms also that genes play a role in the expression of the character and suggests that this aspect might be improved using the DH plant production technique.

Table 3.4. Genetic progress for androgenetic response through recurrent selection

Variables Recurrent cycle	Number of plated anthers	AC	ELS	PLA	F2n	DH %
C0	12.248	13.0%	56	5.0	34.4	0.47
C1	12.719	29.5%	147	11.4	37.3	0.98
C2	24.284	31.5%	217	22.7	40.8	4.62
Genetic progress	-	142%	282%	355%	35%	893%

This might be a very powerful method if we search to integrate recurrent selection for agronomic purposes, and DH plant production through *in vitro* androgenesis. However a large number of genes are certainly involved in the whole regeneration pathway from the androgenetic induction to the obtaining of developed plantlets and their progeny, and selection might increase the frequency of more adapted alleles. There is a need for the evaluation of genetic correlations between anther culture response, which has a real polygenic character, and the other traits to be taken into account for practical breeding.

3.6 Conclusions

Since the original discovery of androgenesis by Gua and Maheshvari in 1964, little progress has been made in elucidating the cellular and molecular mechanisms of androgenetic induction. There is clearly a need for a better understanding of this kind of mechanism in order to address the real problem of recalcitrant genotypes. It would also render this method more easy to use.

Using highly androgenetic responsive material will provide the identification, by molecular techniques, of putative genes or gene products involved in androgenetic induction. The creation and analysis of near isogenic lines for responsiveness on the other hand, might be a powerful method giving some idea on the causal biological functions involved.

However, we have to keep in view, for practical applications in breeding programs, that induction of androgenesis is not the only bottleneck to avoid. The other steps of regeneration such as androgenetic embryo development and germination, spontaneous chromosome doubling, easiness to self the regenerated plants... are also critical steps with genetic controls. It will certainly take some time to elucidate all of these.

Acknowledgments

The research work was supported by the "Institut National de la Recherche Agronomique" and in part by Limagrain/Biocem and Rhone Poulenc Agrochimie. This studies would not have been possible without the expert technical assistance of M. Brinkmann, J. Tucci and S. Antoine-Michard;. Dr Barloy, Dr Dieu, Dr Murigneux provided invaluable contribution to this research. We also acknowledge F. Vear for linguistic revision of the manuscript.

References

Agache S, Bachelier B, de Buyser J, Henry Y, Snape J (1989) Genetic analysis of anther-culture response in wheat using aneuploid, chromosome substitution and translocation lines. Theor Appl Genet 77:7 – 11

Alexander MP. (1969) Differential staining of aborted and non aborted pollen. Stain Technol 44(3):117 – 122

Anonymous (1975) Primary study on induction of pollen plant of Zea mays L. Acta Genet Sin 2:142 - 143

Armstrong CL, Romero-Severson J, Hodges TK (1992) Improved tissue culture response of elite maize inbred through backcross breeding, and identification of chromosomal regions important for regeneration by RFLP analysis. Theor Appl Gene. 84: 755 – 762

Barloy D, Denis L, Beckert M (1989) Comparison of the aptitude for anther culture in some androgenetic doubled haploid maize lines. Maydica 34:303 – 308

Barloy D, Beckert M (1993) Improvement of regeneration ability of androgenetic embryos by early anther transfert in maize. Plant Cell Tissue and Organ Culture 33:45 – 50

Beaumont VH, Rochefort TR, Widholm JM (1995) Mapping the anther culture response genes in maize. Genome 38:968 – 975

Beckert M, Pollacsek M (1979) Expression de la variabilité génétique du maïs en différentes conditions de cultures de tissus. Ann Amélior Plantes 2(5):563 – 581

Becraft PW, Taylor GA (1992) Genetic variation for anther callus inducibility in crosses of highly culturable winter wheats. Plant Breeding 108:19 – 25

Cowen NM, Johnson CD, Armstrong K, Miller M, Woodsley A, Peticelli S, Skoku M, Belmar S, Petolino JF (1992) Mapping genes conditioning in vitro androgenesis in maize using RFLP analysis Theor App. Genet 84:720 – 724

Cummings D, Green CE, Stuthman DD (1976) Callus induction and plant regeneration in oats. Crop Sci 16:465 – 470

Dieu P, Beckert M (1986) Further studies of androgenetic embryo production and regeneration from in vitro cultured anther of maize. Maydica 31: 245 – 259

Duncan DR., Williams ME, Zehr BE, Widholm JM (1985) The production of callus capable of plant regeneration from immature embryos of numerous Zea mays genotypes. Planta 165:322 – 332

Foroughi-Wehr B, Friedt, Wenzel G (1982) On the genetic improvement of androgenetic haploid formation in hordehum vulgare L. Theor Appl Genet 62:233 – 239

Genovesi AD and Collins GB (1982) In vitro production of haploid plants of corn via anther culture. Crop Sci 22:1137 – 1144

Guha-Mukherjee S (1973) Genotypic differences in the in vitro formation of embryoids from rice pollen. J. Exp. Bot. 24 (78):139 – 144

Guha S, Maheshwari SC (1964) In vitro production of embryos from anther of Datura. Nature 204:497

Guiderdoni E (1991) Gametic selection in anther culture of rice (Oryza sativa L.) Theor Appl Genet 81:406 – 412

Hanzel JJ, Miler JP, Brinkmann MA, Fendos E (1985) Genotype and media effects on callus formation and regeneration in Barley. Crop Sci 25:27 – 31

Henry H, Bernard S, Bernard M, Gay G, Marcotte LJ, De Buyser J (1993) Nuclear gametophytic genes from chromosome arm 1RS improve regeneration of wheat microspore-derived embryos. Genome 36:808 – 814

Hodges TK, Kamo KK, Imbrie CW, Becwar MR (1986) Genotype specificity of somatic embryogenesis and regeneration in maize. Biotechnology 4:219 – 223

Ma H, Gu M, Liang G. (1987) Plant regeneration from cultured immature embryos of Sorghum bicolor (L.) Moench. Theor Appl Genet 73: 389 – 394

Miao SH, Kuo CS, Kwei Y, Sun AT, Ku S., Lu WL, Wang Y. (1981) Induction of pollen plant of maize and observations of their progeny. In plant Tissue Culture: Proceeding of Beijing Symposium 1978 Pitmann Publishers, Boston pp 23 – 24

Miha A., Earle ED, Khush GS. (1985) Inheritance of callus formation ability in anther culture of rice (Oryza sativa L). Theor Appl Genet 70:113 – 116

Murigneux A, Baud S, Beckert M (1993) Molecular and morphological evaluation of doubled haploid lines in maize. 2. Comparison with single seed descent lines. Theor Appl Genet 87:278 – 287

Murigneux A, Bentolila S, Hardy T, Baud S, Guitton C, Jullien H, Ben Tahar S, Freyssinet G, Beckert M(1994) Genotypic variation of quantitative trait loci controlling in vitro androgenesis in maize. Genome 37:970 – 976

Nakamura C, Keller WA (1982) Plant regeneration from inflorescence culture of hexaploïd triticale. Plant Science 24: 275 – 280

Petolino JF and Jones AM (1986) Anther culture of elite germplasm of maize. Crop Sci. 26:1072 – 1074

Petolino JF, Thompson SA (1987) Genetic analysis of anther culture response in maize Theor. Appl Genet 74:284 – 286

Petolino JF, Jones AM, Thompson SA (1988) Selection for increased anther culture response in maize Theor Appl Genet 76:157 – 159

Reddy VS, Leelavathi L, Sen SK (1985) Influence of the genotype and culture medium on microspore callus induction and green plant regeneration in anthers of Oryza sativa. Physiol Plant 63:309 – 314

Snape JW, Wright AJ, Simpson E (1984) Methods for estimating gene numbers of quantitative characters using doubled haploid lines. Theor Appl Genet 67:143 – 148

Sears RG, Dekard EL (1982) Tissue culture variability in wheat: Callus induction and Plant regeneration. Crop Sci 22:546 – 550

Thompson DM, Chalmers K, Waugh R, Forster BP, Thomas WTB, Caligari PDS, Powell W (1991) The inheritance of genetic markers in microspore-derived plants of Barley Hordeum vulgare L. Theor Appl Genet 81:487 – 492

Wan Y, Rochefort TR, Widholm JM (1992) RFLP Analyse to identify putative chromosomal regions involved in the anther culture response and callus formation of maize. Theor Appl Genet. 85:360 – 365

4 Experimental Control of Pollen Development

E. Heberle-Bors
Institute of Microbiology and Genetics, University of Vienna, Dr. Bohrgasse 9,
A-1030 Wien, Austria

4.1 Introduction

Pollen development is a tightly regulated process which begins with meiosis, proceeds through microspore development and gives rise to pollen grains which mature. Pollen development can be seen separate from pollen function, i. e. pollination and fertilization which, however, also involve developmental processes such as pollen germination, tube growth, sperm release. Pollen grains represent the gametophytic generation in the alternation of generations in higher plants. Their development takes place inside an organ of the sporophyte, i. e. the anther. Pollen thus depends on the sporophyte in which it develops as an endophyte.

Pollen represents an interesting experimental system for developmental biology. Pollen is the only tissue in higher plants which consists of isolated cells without cell wall connections for the most part of its development. It allows to address questions about the developmental autonomy of the gametophytic generation and the degree of sporophytic control. Pollen development is also an interesting system of peculiar cell cycle phenomena, ranging from meiosis to the asymmetric division of the polarized microspore and developmentally regulated cell cycle arrests in the maturing pollen grain. Finally, isolated microspores cultured *in vitro* have the remarkable property of switching to a sporophytic mode of development after being stressed resulting in the formation of haploid embryos and plants.

4.2 The gametophytic pathway

4.2.1 *In vitro* pollen development

The isolation of male-sterile or gametophyte-lethal mutants is a powerful method to isolate genes controlling pollen development (see the chapter of David Twell in this book). Another approach to study pollen development is to separate pollen from the plant and culture it under controlled conditions *in vitro*. An early success of this approach was that *in vitro* cultured meiocytes were able to pass through much of meiosis indicating that meiosis is quite independent from the surrounding anther tissue (Ito and Stern 1967). The anther wall is however clearly required since the dissolution of the callose wall which surrounds the tetrads requires the action of a tapetum-produced enzyme, a ß-1,3 glucanase.

In vitro culture of isolated microspores has been used to obtain pollen which produced structures resembling pollen tubes (Tanaka and Ito 1980), but never was it possible to obtain truly fertile pollen. It was assumed for some time that microspores do not develop cell-autonomously within the anther but require instruction, possibly by transfer of RNA or proteins from the anther wall to the microspore or young pollen grain (Heslop-Harrison 1980). We have shown for the first time that the microspore is indeed autonomous (Benito Moreno et al. 1988) by production of seeds after *in vivo* pipet pollination with *in vitro* matured pollen. The culture conditions did not provide any instructions concerning the future development of the isolated microspore, the only role being to maintain viability and to provide nutrients. Doubts whether tube-like structures emerging from *in vitro* cultured pollen seen in earlier experiments were really functional pollen tubes, were removed.

Isolated microspores prior to first pollen mitosis required a complex rich medium enriched with amino acids and nucleotides while young bicellular pollen grains could do with a less complex medium (Tupy et al. 1991). Clearly, the *in vitro* conditions simulated the situation *in vivo*. High sucrose concentrations reduced water potential in the *in vitro* maturing pollen as does the drying out of the pollen during *in vivo* development, and proline added to the cultures substituted for proline accumulation in *in vivo* pollen, also probably as an adaptation to low water potential. Both factors indeed increased the quality of the *in vitro* matured pollen, i.e. seed set after pipet pollination.

The earliest stage at which microspores have been used for successful maturation *in vitro* is prior to S phase of the microspore cell cycle (Touraev et al. 1995). At this stage, the microspore is not yet fully developed - the vacuole is small, the microspore nucleus occupies a central position in the spore and the exine is not yet completely formed. It remains to be seen, however, whether still younger microspores, right after the release from the callose wall, can mature *in vitro*.

In vitro maturation differs in a number of interesting aspects from other plant *in vitro* culture systems. Usually plant cells when isolated from the context of their tissue and organ (their position) dedifferentiate and revert to the basic developmental program of any higher organism: embryogenesis. Microspores do not dedifferentiate when isolated but continue on the developmental path they had embarked on when still inside the organ of origin. In this sense microspores are similar to animal cells which maintain their level of differentiation *in vitro* and can be induced to differentiate further. The reason why microspores are so different can be seen in their being true single cells which distinguishes them from any other cell type in the plant body. This lack of tissue integration and autonomy seems to require the necessity for a much stronger cell-autonomy, a truly "wired-in" type of developmental regulation uncommon to other plant cell types. Otherwise pollen development would become to be too dependent upon environmental influences and pollen grains would probably not be able to fulfill their function in reproduction: sterility would be much more frequent under unfavourable environmental conditions.

The second feature which distinguishes *in vitro* maturation from other *in vitro* systems is the lack of hormones in the medium. Mature pollen and anthers do contain growth substances such as auxin, gibberellins, etc., some in surprisingly large amounts (although these estimates have to be seen with care since they are based on out-dated techniques) but nothing is known about their function. Apparently they are not required for pollen development but possibly for pollination.

Originally, seed set after pollination with the *in vitro* matured pollen was very low, even under optimal conditions of *in vitro* maturation leading to a high number of apparently mature pollen. Also germination frequency, determined *in vitro* after transfer of the *in vitro* matured pollen to a germination medium, was low. The *in vitro* pollen seemed to lack something which the *in vivo* pollen had. An ultrastructural investigation confirmed that the *in vitro* pollen looked like normal pollen (Kastler 1990). Apparently the *in vitro* pollen was not really mature and lacked something we could not detect in the electron microscope and which the *in vivo* pollen possibly obtained from the surrounding anther wall. Addition of a pollen diffusate obtained by incubation of mature, dry *in vivo* pollen in a buffer for a few seconds, increased *in vitro* germination frequency and tube growth of the *in vitro* pollen strongly and also improved seed set after pipet pollination. Further analysis showed that the active compounds in this diffusate were flavonols like quercetin or kaempferol which accounted for 80% of the diffusate effect (Ylstra et al. 1992).

Flavonols are known to be present in large quantities in pollen of all higher plants (Wiermann and Viert 1983). They serve various functions: they are anti-oxidants, they protect the germ line against UV-irradiation, and they attract pollinators due to their yellow colour (Koes et al. 1994). We established that flavonols have growth-promoting activity on pollen tubes. During the course of these experiments other labs found that anti-sense suppression of chalcone synthase (CHS), the entry enzyme into the flavonoid biosynthesis pathway, in petunia caused male sterility or better conditional male fertility (van der Meer et al. 1992, Taylor and Jorgensen 1992). (The difference in the two terms is significant. Pollen of the transgenic plants was non-functional only in selfings but fertile in cross-pollinations and the pollen of these plants looked perfectly normal under the microscope. Sterility does not depend on the type of cross and sterile pollen usually is arrested at a particular stage of development.) The anti-sense CHS-work did not provide a clue about which flavonoids were causing the developmental arrest while our work strongly suggested that it is the flavonols which were lacking in pollen of the CHS-suppressed plants. This was confirmed lateron (Mo et al. 1992). Furthermore, the flavonoid biosynthesis genes were found to be active only in the anther wall tissues but not in the pollen (Koes et al. 1990). This allowed us to conclude that our *in vitro* pollen was indeed mature but that it lacked for its further function flavonols which are produced by "paternal", i. e. sporophytc, tissues and are stored somewhere in the pollen to be used after landing of the pollen on the stigma and hydration.

We then routinely included flavonols or pollen diffusate into the culture medium used for pipet pollination, and - with proper handling - achieved seed set in the range of normal *in vivo* pollination (about 1000 seeds per pod in SR1 tobacco). Recently we have omitted these compounds after it has been discovered that wounding of flowers induces flavonol formation in the stigma and style (Pollack et al. 1993, Touraev et al. 1997). The stigma-produced flavonols can sufficiently compensate for the lack of pollen-derived flavonols.

Another class of growth regulators which we detected to affect pollen tube growth were steroids. Various animal and plant steroids were added to *in vitro* matured pollen, and testosteron but also oestrogen stimulated pollen germination and pollen tube growth (Ylstra et al. 1995). In combination with flavonols an additive effect was found which indicated that the two classes of growth substances act independently. Plant steroids such as brassinosteroids had either no or an inhibitory effect. Possibly their function is to block precocious pollen germination in the anther.

The effect on pollen germination and tube growth seen with flavonols and steroids is paradigmatic for the heuristic value of *in vitro* matured pollen. Their lack of contact with the anther wall deprives them of substances with which the sporophyte normally impregnates the pollen to ensure its reproductive success. *In vitro* pollen can thus be used to screen for substances affecting reproductive effectiveness of pollen, be it in a positive and possibly also in a negative sense. It remains to be seen whether *in vitro* pollen offers an advantage over normal pollen in *in vitro* germination assays to test for environmental pollutants.

Sporophytic control of pollen development is not only seen in providing flavonols or other substances required for proper pollen function. The sporophyte can control reproduction also via restricting nutrient flow to the developing pollen (pollen sterility) or by interfering with recognition phenomena between pollen and stigma/style (incompatibility/incongruity). In all these cases, *in vitro* maturation may be used to produce fertile pollen which can be used for "self"-pollination to produce homozygous lines of the respective sterility or incompatibility gene. In male sterility, obviously *in vitro* culture can only be used in those cases in which male reproductive development in arrested after meiosis but there should be no reason why a microspore isolated from an aborting anther should not be rescued by *in vitro* culture. However, no example has been published to prove the feasibility of this approach.

4.2.2 Pollen selection

Ever since the rediscovery of the Mendelian laws it has been found in a number of traits that segregation in the offspring can sometimes deviate substantially from the expected ratio. Gametic selection is one of the mechanisms which have been evoqued to explain these deviations. In plants, there exists an old controversy about the reality of gametic selection and quite a body of evidence has been obtained to substantiate its existence in plants (Ottaviano et al. 1986). Most of the evidence has been provided by genetic, i. e. statistical, analysis after applying selection pressures to whole plants or male inflorescences (Sari-Gorla et al. 1986). The problem with this kind of approach is that the sporophytic tissues surrounding the pollen protects the pollen and buffers any selective agent applied to a whole male inflorescence, male flower or anther. It is also intrinsicly difficult to distinguish between direct effect on the pollen and indirect effects which primarily affect the surrounding sporophytic tissues.

We set out to directly prove the feasibility of pollen selection by applying selection pressure to pollen developing *in vitro* and using this pollen for *in vivo* pipet pollination. A transgenic reconstruction approach was used by producing transgenic tobacco plants via conventional Agrobacterium-mediated transformation which contained, in a single insert, a construct consisting of the hygromycin resistance gene behind a strong pollen-expressed promoter and the kanamycin resistance gene behind a sporophyte-expressed promoter (Touraev et al. 1995). The pollen-expressed promoter used was the DC3-promoter of a carrot *lea* gene which is induced during initiation of somatic embryogenesis. This promoter turned out to be by far the strongest promoter during pollen development, as compared to the LAT52- or other promoters. Isolation of microspores and their culture in a hygromycin-containing medium produced 50% *in vitro* matured pollen which, after pipet pollination of wild-type plants, produced 100% kanamycin-resistant seeds. A single day of hygromycin application was sufficient to efficiently select the resistant pollen. Such a short period for selection is unrecorded in plant genetics.

This was direct proof that pollen can be selected during pollen development. Given the large number of pollen grains and their haploidy, application of a selection pressure on *in vitro* maturing pollen may become a means to increase the frequency of selectable genes in the next generation, at least for genes which are expressed both in the gametophytic and the sporophytic generation.

This latter point, however, is not trivial. Originally, it was thought that there exists a large overlap of genes being expressed in both generations, based in part on RNA-RNA hybridization experiments (for a review see Mascarenhas 1990). However, in the last years genes a number of cases have been reported in which gene families consist of sporophytic and gametophytic members. Examples are the actin and the alcohol dehydrogenase gene families. Other examples are the Betv1 and Betv2 gene families of birch (see below).

RNA-RNA hybridizations cannot discriminate between RNAs being transcribed from closely related genes of a gene family. Thus, it is possible that the extent of a sporophyte-gametophyte overlap has been overestimated, and selection acting on a gametophytic member of a gene family will not lead to a enhanced performance in the succeeding sporophytic generation. Some genes, however, do exist which seem to be expressed in both generations. Some embryo-lethal mutants have been shown to have a gametophytic effect and are likely to be candidates to isolate genes for a gametophytic-sporophytic overlap (Ottaviano et al. 1987).

4.2.3 Male germ line transformation

Having developed an efficient protocol for the development of fertile pollen *in vitro*, we hypothesized that transfer of DNA into microspores would deliver this DNA into the egg cell. Unlike other proposed methods of pollen transformation, there was the clear rational that a microspore represents a germ line cell for the male gametes so that DNA introduced in the germ line cell would be present also in the gametes. Attempts to deliver DNA to mature pollen or just adding DNA to stigmas during pollination always faced the problem of how the DNA would enter the gametes (Heberle-Bors et al. 1990, Stöger et al. 1992). Our method consisted of transferring DNA into microspores, allowing them to develop into mature pollen *in vitro* and to use the *in vitro* matured pollen for *in vivo* pipet pollination in order to produce seeds from which transgenic seedlings could be selected.

We indeed succeeded to produce transgenic tobacco plants via this method which we called male germ line transformation (MAGELITR). Microspores were isolated in the G1/S phase of the cell cycle and bombarded with gold particles coated with DNA encoding the NPT II gene as a marker to be used for seedling selection on kanamycin. Five seedlings could be selected which turned out to be transgenic, as just by a complete Southern blot analysis which confirmed physical integration of the transferred DNA into high molecular weight DNA of the recipient tobacco plant (Touraev et al. 1997). Genetic analysis confirmed stable transmission of the transgenes to the two next generations. Two different plasmids had been used in the transformation experiments with plasmid-specific marker genes, and indeed four plants were recovered which contained the marker present in the DNA used for transformation while one transgenic plant was produced which contained the marker specific to the other plasmid used.

We called this method male germ line transformation because it consists of transferring DNA not into male gametes (the sperm cells) but into a cell which is a progenitor of the sperm cells. Pollen mother cells, microspores and the generative cells of pollen grains constitute a short germ line which ends in the production of the male gametes. DNA transferred in any of these cells will result in the formation of male gametes which transfer their DNA - with the help of the vegetative cell of the pollen grain and the female sex organ - into egg cell and polar cell.

MAGELITR is a regeneration-independent method for transformation of sexual plants. It avoids chimerism due to the single-celled nature of the explant, and should also avoid somaclonal variation which often is the result of lengthy *in vitro* culture procedures. MAGELITR is simple: *in vitro* maturation in tobacco requires only one week, the rest is done by the pollen recipient plant. Only two cell cycles take place during this period, in a developmental program which is identical to the situation *in vivo*. In all other transformation methods, in contrast, many cell generations take place *in vitro* under sometimes quite uncontrolled conditions which are good breeding grounds for mutations.

MAGELITR is also fast: one week of *in vitro* culture, three weeks for the formation of seeds and the selection of a transgenic plant further 2 weeks, in total 6 weeks. In addition, MAGELITR might be genotype-independent. We have produced mature pollen from microspores with any cultivar we have tried and also in wheat, the genotype had no effect on *in vitro* maturation (Stauffer et al. 1991).

Time will tell whether there is a need for MAGELITR as an alternative to other methods of transformation. The major crop plants have all been transformed with other methods. Only when any of the mentioned advantages and differences become relevant in the future, may MAGELITR become a more generally used transformant technique. Some species do exist which can not or only very inefficiently be transformed by any of the conventional methods, f. e. *Antirrhinum majus* or *Capsicum annuum*, which represent obvious candidates for a MAGELITR approach.

4.2.4 Pollen allergy

An important aspect of "haploidy" which is not directly related to "experimental control of pollen development" is the property of pollen to cause allergies in humans. To put it very simply, clinicians are very interested and active in cloning pollen allergy genes. This research provides the "haploidist" with pollen-expressed genes.

We have collaborated with colleagues of the Vienna General Hospital in cloning and analyzing allergy genes of birch (*Betula verucosa*). Betv1 is the major pollen allergen of birch (more than 80% of persons allergenic to birch pollen produce IgE antibodies against Betv1; Betv1 comprises more than 20% of total protein of birch pollen). After cloning the Betv1 gene (Breiteneder et al. 1989), it turned out that it showed homology to PR-protein genes, i.e. to PR-1 of parsley (Somssich et al. 1986). Another surprise was that Betv1 constitutes a quite large gene family in birch (with a genome size not much bigger than *Arabidopsis thaliana*). We identified in pollen more than 10 Betv1 genes (Swoboda et al. 1994, 1995a), and three more from somatic tissues (Swoboda et al. 1995b). In situ hybridisation revealed that the pollen-expressed members of the gene family are pollen-specific, i.e. they are expressed late in pollen development but are not expressed in other tissues, also not in sporophytic tissues of the anther (Pettenburger et al. 1993, Swoboda et al. 1995c). In contrast, the three somatic members are normally quiescent but are transcriptionally induced by pathogens (Swoboda et al. 1995b). Finally, it turned out that Betv1 seems to be a ribonuclease (Swoboda et al. 1996) which left us with the question what a ribunuclease does as a PR-protein, and more specifically what a ribonuclease does in pollen. In fact it seems that Betv1 does not do anything in pollen because upon rehydration it is quantitatively washed out from the pollen. Therefore we assume a function on the stigma, during pollination. Intriguingly, ribonucleases are involved in gametophytic self-incompatibility, however, as style-expressed genes.

The Betv1 genes are also interesting in the context of haploid-specific genes, as discussed in the chapter on pollen selection. The Betv1 gene family is a good example of how higher plants control expression of genes which have specific functions in various phases throughout the life cycle. Rather than using alternative splicing, plants seem to use the "institution" of the gene family as a means to put the expression of genes under developmental and/or environmental control. Therefore, Betv1 would not be a good candidate for pollen selection since selection on a pollen-specific member of the gene family will not increase the frequency of a somatic member in the next generation. Only if the individual genes of the family are located close to each other on the same chromosome - as the result of a recent gene duplication -, linked transmission will occur.

Another birch allergen is Betv2. It shows homology to profilin, an actin-binding protein in eucaryotes (Valenta et al. 1991). Profilin is present in mature, dry pollen in a profilactin complex which dissociates upon rehydration to release actin monomers for the formation of the microfilament system involved in pollen germination and tube growth (Valenta et al. 1993). In tobacco, there exists at least one gene which seems to be expressed only in pollen (Mittermann et al. 1995, 1996).

4.3 The sporophytic pathway

4.3.1 Stress-induced microspore embryogenesis

Organ cultures of whole anthers at the microspore stage were once thought to contribute to an understanding of pollen development but it was somewhat of a surprise when it was discovered that what "hatched" from the anther locules after some time of *in vitro* culture were not mature pollen grains but embryos (Guha and Maheshwari 1964) which turned out to have the gametic (haploid) number of chromosomes. These embryos developed into haploid plants which after colchicine treatment were diploid and homozygous. The potential of this technique for genetics and plant breeding was quickly recognized, and efforts to investigate microspore embryogenesis for doubled haploid production were undertaken on a world-wide level.

Originally discovered in *Datura innoxia*, tobacco was the first species in which this unexpected phenomenon was confirmed. Jean-Pierre and Colette Nitsch and Jean-Pierre Bourgin were the initial driving forces, together with Norman Sunderland in the UK, who investigated how general microspore embryogenesis was in higher plants and which factors controlled this deviation from normal pollen development. Later, French investigators around Demarly such as Yves Henry, Jacques de Buyser and Emmanuel Picard opened another frontier when they succeeded to produce doubled haploid from anther cultures of the first cereal (wheat) and Chinese scientists soon followed suit to make anther cultures a so-called "science for the people" during the so-called "Cultural Revolution".

Anther culture remained the technique of the day for quite some time although it has been recognized early on that only microspore cultures would allow to analyze microspore embryogenesis properly and that only this technique would fulfill the dream of the geneticist and breeder to produce a large population of doubled haploids from a cross which would represent the genetic variability produced by male meiosis (Nitsch and Norreel 1973, Reinert et al. 1975). Anther culture was simple and reproducible while pollen or microspore culture was highly unreproducable and inefficient for many years. The breakthrough came when it was recognized that one single factor was essential to induce embryogenesis in microspores or young pollen grains: stress.

This breakthrough was, however, proceeded by small steps in the improvement of microspore cultures. For example, it was found that a simple culture medium is sufficient for embryogenesis, without growth substances, and that one needs high sugar concentrations. Other findings were that FeEDTA was required (Nitsch 1969), or that an elevated pH is needed as compared to other *in vitro* systems (Heberle-Bors 1985 for a review). The most important finding was that some kind of pretreatment very beneficial. Cultured anthers without pretreatment did produce plantlets or at least some aborted embryos in many species while pretreatments increased plantlet numbers, sometimes considerably.

In the early days these pretreatments were left to the predilection of the researcher and were performed at all levels conceivable: they were done on the whole plant level, on excised inflorescences, on excised flowers, on excised and *in vitro* cultures anthers and finally also on isolated and *in vitro* cultured microspores (for a recent review see Touraev et al. 1997). Other *in vitro* regeneration systems did not require such elaborate pretreatments. Isolation from the mother plant and the right combination of auxin and cytokinin was considered sufficient. Apparently something was different with microspore cultures.

This difference became obvious when true microspore cultures were performed which did not include any pretreatment prior to culture. Without stress the microspores developed into apparently normal pollen (see chapter on *in vitro* pollen development). Only when tobacco pollen cultures were stressed by a starvation treatment (Kyo and Harada 1986) or *Brassica napus* microspore cultures were heat-stressed (Custers et al. 1994) were the microspores reprogrammed to develop into an embyro. No pretreatments of anthers or flower buds were used in these experiments thus excluding any effect of the anther wall or the donor plants. Clearly, the stress treatments were necessary and sufficient to induce embryogenesis.

But why were there different stress treatments for the two species and why was it necessary to use microspores in rape seed but bicellular pollen grains in tobacco?

In fact, a closer look revealed that this was not a question of principle but of expertise. Also in tobacco, microspores can now be induced to develop into microspores, and in microspores a heat shock is also effective (Touraev et al. 1996a). A combined treatment has even an additive effect.

These results provided the basis for a concept of a universal mechanism to induce microspore embryogenesis. According to this concept, stress can induce a microspore or immature pollen grain at any stage of development, or more precise, at any phase of the cell cycle (Touraev et al. 1996b).

Flow cytometry and thymidine labeling of replicating DNA revealed that microspores in the G1 phase of the cell cycle pass through S phase during the stress treatment and arrest in the G2 phase of the cell cycle (Touraev et al. 1996b). Only after transfer to a rich medium at lower temperature do the induced microspores undergo mitosis and continue to divide. Microspores in the G2 phase divide during the stress treatment and become arrested in G1. Finally, vegetative cells which *in vivo* are arrested in G1, undergo DNA replication during the stress treatment and are - as G1 microspores - arrested in the G2 phase (Zarsky et al. 1992). The generative cell is not affected by the stress. Thus, the second aspect of induction is a cell cycle arrest, at least in tobacco. In *B. napus*, however, no cell cycle arrest is observed indicating that possibly the starvation treatment is responsible for the cell cycle arrest.

The third aspect is that after transfer to a rich medium, the induced microspores overcome the cell cycle arrest and divide to form eventually an embryo. Thus, induction of microspore embryogenesis comprises a stress-induced cell cycle arrest not seen during normal pollen development which after stress release leads to sustained cell division while the specific, developmentally regulated cell cycle arrest during normal pollen development is derepressed.

Most of the stressed microspores divide by a regular asymmetric division while a fraction divide symmetrically to form two similar-sized cells. By adding an extract from whole anthers it was possible to induce symmetric divisions of the *in vitro* cultured microspores. Under maturation conditions, these microspores developed into mature pollen grains from which one of the two similar-sized cells produced a pollen tube (Touraev et al. 1995). This indicated that the symmetry of first pollen mitosis is not essential for determining the developmental fate of the pollen. Similar results have been obtained with colchicine-treated microspores (Eady et al. 1995).

Possibly more important than the question of symmetry is that the induced microspores are irreversibly committed to embryogenesis. Transfer of tobacco microspores which had undergone a starvation and heat shock treatment, to a maturation medium which contains a higher sucrose concentration than the embryogenesis medium, as well as amino acids and nucleosides, did not lead to a continuation of gametophytic pollen development but to a few cell divisions followed by degeneration (Touraev et al. 1996b). Thus, the stress treatment not simply arrested pollen development but really reprogrammed them towards sporophytic type of development. Apparently changes in gene expression must have happened *during* the stress treatment which were responsible for the developmental switch. Two-dimensional gel electrophoresis of proteins did not detect any differences between embryogenic and non-embryogenic microspores (Garrido et al. 1993). Only after *in vitro* translation of isolated mRNA was it seen that indeed new genes are expressed during the stress treatment which are not translated. One of these genes apparently is a 18kDa heat shock protein gene which is induced in starvation-induced microspores under normal temperature (Zarsky et al. 1995).

An alternative explanation for embryogenic induction in microspores could be that the stress treatment leads to an accumulation of stress-induced proteins which would affect gene expression and thus pollen development *after* transfer to non-stress conditions. Indeed, changes in the levels of phosphorylated proteins have been detected in stressed microspores (Kyo and Harada 1980). In addition, we could detect changes in protein kinase activities during the inductive stress treatment (Garrido et al. 1993).

At present a number of labs including our own are active in cloning genes specifically involved in embryogenic induction of microspores. Some labs are using differential display of reverse transcribed mRNAs to isolate stage-specific genes, others are using a genetic approach to identify QTLs contributing to microspore embryogenesis by comparing well-responding with non-responding lines, f. e. in maize. A mutant approach should be possible in maize but is not feasible in *Arabidopsis thaliana* since microspore culture in this species is not yet established and a mutant screen by microspore culture would be very laborious.

The reprogramming of microspores by stress and the concomittant cell cycle arrest has an interesting parallel in mammals. For decades it had been assumed that adult mammalian cells cannot be reprogrammed to embryogenesis, that in fact adult mammalian cells may not be totipotent. Also in frog, transplantation of nuclei from adult cells into enucleated egg cells did not result in full development. Only larvae were formed. Only recently has it been shown in sheep that serum withdrawal from *in vitro* cultured primary mammary gland cells leading to a cell cycle arrest was required to reprogramme these cells towards embryogenesis after fusion with enucleated egg cells (Wilmut et al. 1997). Further above we had argued that microspores are different from other plant cells and more similar to animal cells in the sense that they maintain their original developmental orientation *in vitro* . Now a further parallel is seen in that a stress is required in microspores and animal cells to release their totipotency and to reprogramme them for embryogenesis.

4.3.2 Microspore culture in wheat

Doubled haploid production (DH) in wheat was one of the first applied successes of anther culture. The first variety of a cereal bred with the DH method was "Florin" in France (Henry and de Buyser 1990). However, despite much progress in anther culture technology, increases in DH production were not really spectacular, particularly when seen on the background of its potential which is the number of microspores produced by an anther. In wheat one anther contains 2000 to 3000 microspores while plantlet yield per anther is around 1.

Having set-up a microspore culture system in tobacco which allowed the induction of basically all living microspores to undergo sustained cell divisions and the production of large amounts of torpedo-shaped embryos, we were interested to find out whether the results obtained with tobacco could be directly applied to wheat. Wheat anthers were isolated and stressed by heat and starvation in the same medium as used for tobacco. Microspores were isolated from these anthers and were combined with microspores which shed spontaneously from the stressed anthers. Density centrifugation in a Percoll gradient allowed to separate the living from the dead microspores. In a rich medium conditioned with ovaries, more than 50% of the microspores underwent repeated cell divisions, and high numbers of plantlets were produced in all wheat genotypes tested so far, even in genotypes which were recalcitrant in anther culture (Touraev et al. 1996c, and unpublished results).

These results represented a major breakthrough in DH production. For the first time was it possible to produce large amounts of embryos in this species and to efficiently eliminate the genotypic limitation in DH production which hampered doubled haploid breeding in this species for so long. This result could be a turning point in DH breeding. Now it will be possible to make full use of meiotic recombination by raising large numbers of DHs from a F1-hybrid and to screen them for the desired combination of traits.

4.3.3 Transformation of embryogenic microspores

Not only microspores developing *in vitro* to mature pollen have been used for transformation but also embryogenic microspores (Stöger et al. 1995, Alwen et al. 1992). We have targeted young bicellular microspores biolistically after the inductive starvation treatment when they were already for a few days in the rich medium. The DNA used coded for the ß-glucuronidase gene behind the 35S-promoter together with the NPT II gene behind the nos-promoter. Kanamycin selection turned out to be unsuccessful. Fortunately at that time a novel, non-toxic GUS substrate, IMAGENE GREEN (Molecular Probes Inc.) became available which we used to select globular to heart-shaped embryos under the fluorescent microscope. With a pipet we picked green-fluorescing embryos and transferred them to a less rich agar medium for further development. We were able to transfer hundreds of GUS-positive embryos from a few Petri dishes which represented one inoculation of microspores. We also bombarded other tissues with the 35S-GUS-gene. However, autofluorescence and non-specific staining precluded the use of the GUS-gene with IMAGENE GREEN as a vital marker gene. Good results were obtained only with microspore embryos.

Unfortunately we did not have a technology in place which would have allowed us to grow most of the embryos into seedlings. Most of them died on the agar medium. In the end two plants could be raised which appeared to be transformed. Using flow cytometry they turned out to be haploid. Colchicine treatment of lateral buds was successful with one bud in one of the plants and resulted in the production of fertile gametes and seeds. All these seeds germinated on kanamycin indicating that the donor branch was homozygous. A Southern blot analysis confirmed integration of the transferred DNA into the chromosomes of the two haploid plants and the one homozygous DH, and also confirmed the stable transmission to the offspring.

These results provided an alternative approach for transformation in those species in which regeneration from other tissues is impossible or difficult. Particularly in cereals microspores represent the cell type with the highest regeneration potential. Once genotype-independent DH production has been established in a given species and lab, DNA transfer into microspores will easily complement the lab and lead to the production of transgenic plants in any given genotype the breeder is interested in.

Transformation via embryogenic microspores will, however, probably not be the most important application of DHs in molecular breeding. Rather, DHs will be used to rapidly transfer transgenes from a transformed line to a cultivar. In many species, only one or a small number of model genotypes can be transformed. In a typical situation, transgenic lines are selected from a pool of lines transformed with the same gene(s) in this model genotype in a lengthy procedure which includes release experiments. The lines have to be tested for copy number, stability of transmission and stability of expression at a desired level. Rather than trying to transfer the same gene in a target genotype (cultivar, breeding line) of the same species, it may be easier to start from a transformed line of the model genotype and transfer the gene to a target genotype by DH breeding. The testing of a new transgenic line would not be necessary. Clearly, this approach depends on the ease of transformation of target genotypes, of the genetic distance between selected transgenic line and target line and of the regulatory burden for release experiments.

References

Alwen A, Benito Moreno RM, Vicente O, Heberle-Bors E (1992) Plant endogenous ß-glucuronidase activity: how to avoid interference with the use of the E. coli GUS as a reporter gene in transgenic plants. Transgenic Research 1:63 – 70

Benito Moreno RM, Macke F, Alwen A, Heberle-Bors E (1988) In situ seed production after pollination with in vitro matured, isolated tobacco pollen. Planta 176:145 – 148

Breiteneder H, Pettenburger K, Bito A, Valenta R, Kraft D, Rumpold H, Scheiner O, Breitenbach M (1989) The gene coding for the major birch pollen allergen Betv1 is highly homologous to a pea disease resistance response gene. EMBO J 8:1935 – 1938

Custers JBM et al. (1994) Temperature controls both gametophytic and sporophytic development in microspore cultures of Brassica napus. Plant Cell Rep 13:267 – 271.

Eady C, Lindsey K, Twell D (1995) The significance of microspore divisions and division symmetry for vegetative-cell specific transcription and generative cell differentiation. Plant Cell 7:65 – 74

Garrido D, Eller N, Heberle-Bors E, Vicente O (1993) De novo transcription of specific mRNAs during the induction of tobacco pollen embryogenesis. Sex Plant Reprod 6: 40 – 45

Guha S, Maheshwari SC (1964) In vitro production of embryos from anthers of Datura. Nature 204:497

Heberle-Bors E (1985) In vitro haploid formation from pollen: a critical review. Theor Appl Genet 71: 361 – 374

Heberle-Bors E, Benito Moreno RM, Alwen E, Stöger E, Vicente O (1990) Transformation of pollen. In: Nijkamp HJJ, Van Der Plas LHW, Van Aartrijk J (eds) Progress in Plant Cellular and Molecular Biology, Kluwer pp 244 – 251

Henry Y, De Buyser J, (1990) Wheat anther culture: agronomic performance of doubled haploid lines and the release of a new variety 'Florin'. In: Y.P.S. Bajaj (ed) Biotechnology in Agriculture and Forestry Vol 13 Springer, Berlin, Heidelberg

Heslop-Harrison J (1980) The forgotten generation: some thoughts on the genetics and physiology of angiosperm gametophytes. In: Davies DR, Hopwood DA (eds) The plant genome, The John Innes Charity, Norwich, pp1 – 14

Ito M, Stern H (1967) Studies of meiosis in vitro. I. In vitro culture of meiotic cells. Dev Bio, 16:26 – 5

Kastler M (1990) Ultrastrukturelle Unterschiede zwischen in-vivo und in-vitro gereiften Tabakpollen. Diploma Thesis, University of Vienna

Koes RE., Van Blokland R, Quattrochio F, Van Tunen AJ, Mol JNM (1990) Chalcone synthase promoters in Petunia hybrida are active in pigmented and unpigmented cell types. Plant Cell 2:379 – 392

Koes RE, Quattrocio ., Mol JNM (1994) The flavonoid biosynthesis pathway in plants: function and evolution. BioEssays 16:123 – 132

Kyo M, Harada H (1986) Control of the developmental pathway of tobacco pollen in vitro. Planta 168: 427 – 432

Kyo M, Harada H (1990) Specific phosphoproteins in the initial period of tobacco pollen embryogenesis. Planta 182:58 – 63

Löschenberger F, Hänsel H, Heberle-Bors E (1993) Plant formation from pollen of tetraploid wheat species and Triticum tauschii. Cereal Res Commun , 21:133 – 139

Mascarenhas JP (1990) Gene activity during pollen development. Ann. Rev Plant Physiol Plant Mol Biol 41:317 – 338

Meer Van Der IM, Stam ME, Van Tunen AJ, Mol JNM, Stuitje AR (1992) Inhibition of flavonoid biosynthesis in petunia anthers by antisense approach results in male sterility. Plant Cell 4:253 – 262

Mittermann I., Heberle-Bors E., Valenta R (1996) Molecular characterization of profilin isoforms from tobacco (Nicotiana tabacum) pollen. Sex Plant Reprod 9:133 – 139

Mittermann I, Swoboda I, Pierson E, Eller N, Kraft D, Valenta R, Heberle-Bors E (1995) cDNA cloning and characterization of profilin from tobacco (Nicotiana tabacum): increased profilin expression during pollen maturation. Plant Molec Biol, 27:137 – 146

Mo Y, Nagel C, Taylor LP (1992) Biochemical complementation of chalcone synthase mutants defines a role for flavonols in functional pollen. Proc Natl Acad Sci USA 89:7213 – 7217

Nitsch JP (1969) Experimental androgenesis in Nicotiana. Phytomorphology 19:389 – 404

Nitsch C, Norreel B (1973) Effet d'un choc thermique sur le pouvoir embryogène du pollen de Datura innoxia cultivé dans l'anthère ou isolé de l'anthère. C R Acad Sci (Paris) 276, série D, 303 – 306

Ottaviano E, Sari-Gorla M, Frova C, Pe ME (1986) In: Mulcahy, D, Bergamini Mulcahy G, Ottaviano E (eds) Biotcchnology and ecology of pollen. Springer Verlag, New York, pp35 – 42

Ottaviano E, Petroni D, Pe ME (1987) Gametophytic expression of genes controlling endosperm development in maize. Theor Appl Genet 75:252 – 258

Pettenburger K, Tejkl M, Valenta R, Heberle-Bors E, Breitenbach M, Rumpold H, Kraft D, Scheiner O (1993) Distribution of allergens and mRNAs coding for allergens isolated from various tissues of white birch. Molec Immunol 29:1401 – 1406

Pollack PE, Vogt T, Mo Y, Taylor LD (1993) Chalcone synthase and flavonol accumulation in stigmas and anthers of Petunia hybrida. Plant Physiol 102:925 – 932

Reinert J, Bajaj YPS, Heberle E (1975) Induction of haploid tobacco plants from isolated pollen. Protoplasma 84:191 – 196

Sari-Gorla M, Frova C, Binelli G, Ottaviano E (1986) The extent of gametophytic-sporophytic gene expression in maize. Theor Appl Genet 72:42 – 47

Somssich IE, Schmelzer E, Kawalleck P, Hahlbrock K (1986) Gene structure and in situ transcript localization pathogenesis-related protein 1 in parsley. Molec Gen Genetics 213:9398

Stauffer C, Benito Morene RM, Heberle-Bors E (1991) In situ pollination with in vitro matured pollen of Triticum aestivum. Theor Appl Genet 81:576 – 580

Stöger E, Fink C, Pfosser M, Heberle-bors E (1995) Plant transformation by particle bombardment of embryogenic pollen. Plant Cell Rep 14:273 – 278

Stöger E, Benito Moreno RM, Ylstra B, Vicente O, Heberle-Bors E (1992) Comparison of different techniques for gene transfer into mature and immature tobacco pollen. Transgenic Research 1:71 – 78

Swoboda I, Ha DTC, Heberle-Bors E, Vicente O (1995) Expression of Bet v1, the major birch pollen allergen, during anther development: an in situ hybridisation study. Protoplasma 187:103 – 110

Swoboda I, Jilek A, Ferreira F, Hoffmann-Sommergruber K, Scheiner O, Kraft D, Breiteneder H, Pittenauer E, Schmid E, Vicente O, Heberle-Bors E, Ahorn H, Breitenbach M (1995) Isoforms of Bet v 1, the major birch pollen allergen, analyzed by liquid chromatography, mass spectrometry and cDNA cloning. J Biol Chem 270M2607 – 2613

Swoboda I, Scheiner O, Heberle-Bors E, Vicente O (1995) cDNA cloning and characterization of three birch genes, members of the Bet v 1 gene family of pollen allergens, that encode pathogenesis-related proteins. Plant, Cell and Environment 18:865 – 874

Swoboda I, Scheiner O, Kraft D, Breitenbach M, Heberle-Bors E, Vicente O (1994) A birch gene family encoding pollen allergens and pathogenesis-elated proteins. Biochim Biophys Acta 1219:457 – 464

Swoboda I, Hoffmann-Sommergruber K, O'Riordain G, Scheiner O, Heberle-Bors E, Vicente O (1996) Bet v 1 proteins, the major birch pollen allergens and members of a family of conserved pathogenesis-related proteins, are ribonucleases. Physiol Plant 96:433 – 438

Tanaka I, Ito M (1980) Induction of typical cell division in isolated microspores of Lilium longiflorum and Tulipa gesneriana. Plant Sci Let 17:279 – 285

Taylor LP, Jorgensen R (1992) Conditional male fertility in chalcone synthase-deficient petunia. J Heredity 83:11 – 17

Touraev A, Fink CS, Stöger E, Heberle-Bors E (1995) Pollen selection: a transgenic approach. Proc Natl Acad Sci USA, 92:12165 – 12169

Touraev A, Fink CS, Stöger E, Heberle-Bors E (1995) Maintenance of gametophytic development after symmetrical division in tobacco microspore culture. Sex Plant Reprod 8:70 – 76

Touraev A, Fink CS, Stöger E, Heberle-Bors E (1996) Stress-induced microspore embryogenesis in tobacco: an optimized system for molecular studies. Plant Cell Reports 15:561 – 565

Touraev A, Indrianto A, Wratschko I, Vicente O, Heberle-Bors E (1996) Starvation and heat-shock-induced in vitro microspore embryogenesis in wheat (Triticum aestivum L.). Sex Plant Reprod 9:209 – 215

Touraev A, Pfosser M, Vicente O, Heberle-Bors E (1996) Stress as the major signal controlling the developmental fate of tobacco microspores: towards a unified model of induction of microspore/pollen embryogenesis. Planta 200:144 – 152

Touraev A, Stöger E, Voronin V, Heberle-Bors E (1997) Plant male germ line transformation. Plant J, in press

Touraev A, Vicente O, Heberle-Bors E (1997) Induction of microspore embryogenesis by stress. Trends in Plant Sci, in press

Valenta R, Duchene M, Pettenburger K, Sillaber C, Valent P, Bettelheim P, Breitenbach M, Rumpold H, Kraft D, Scheiner O (1991) Identification of profilin as a novel pollen allergen; IgE autoreactivity in sensitized individuals. Science 253:557 – 560

Valenta R, Ferreira F, Grote M, Swoboda I, Vrtala S, Duchene M, Deviller P, Meagher B, Heberle-Bors E, Kraft E, Scheiner O (1993) Identification of profilin as an actin-binding protein in higher plants. J Biol Chem 268:22777 – 22781

Wiermann R, Vieth K (1983) Outer pollen wall, an important accumulation site for flavonoids. Protoplasma 118:230 – 233

Wilmut I et al. (1997) Viable offspring derived from fetal and adult mammalian cells. Nature 385:810 – 813

Ylstra B, Touraev A, Benito Moreno RM, Stöger E, Van Tunen AA, Vicente O, Mol JNM, Heberle-Bors E (1992) Flavonols stimulate development, germination and tube growth of tobacco pollen. Plant Physiol 100:902 – 907

Ylstra B, Touraev A, Brinkmann AO, Heberle-Bors E, Van Tunen AJ (1995) Steroid sex hormones stimulate germination and tube growth of in vitro matured tobacco pollen. Plant Physiol 107:639 – 643

Zarsky V, Garridio D, Rihova L, Tupy J. Vicente O, Heberle-Bors E (1992) Derepression of the cell cycle by starvation is involved in induction of tobacco pollen embryogenesis. Sex Plant Reprod 5:189 – 194

Zarsky V, Garrido D, Eller N, Tupy J, Vicente O, Schöffl F, Heberle Bors E(1995) The expression of a small heat shock gene is activated during induction of tobacco pollen embryogenesis by starvation. Plant, Cell and Environment 18:139 – 147

5 Microspore Culture: a Model for Investigating the Role of Stress in the Induction of Embryogenesis

J.H.G. Cordewener, J.B.M. Custers and M.M. Van Lookeren Campagne[1]
Department of Developmental Biology, Centre for Plant Breeding and Reproduction Research (CPRO-DLO), P.O. Box 16, NL-6700 AA Wageningen, The Netherlands

5.1 Introduction

Embryogenesis is normally initiated by fertilization and the formation of a zygote. Due to the inaccessibility of the egg cell in angiosperms, it is difficult to study the molecular events that take place during fertilization and early embryogenesis *in vivo*, and only very recently have these processes come under experimental control *in vitro* (Faure et al., 1994; Kranz and Lorz, 1993). Progress is being made with genetic analysis of pattern formation in *Arabidopsis thaliana* and maize embryos (Clark and Sheridan, 1991; Jürgens et al., 1991; Laux and Jürgens, 1994; Meinke, 1991), but no clue as yet exists about the mechanism by which embryogenesis and pattern formation are controlled *in vivo*. Fortunately, plants are able to regenerate embryos from somatic cells or tissues, and this somatic embryogenesis has been studied in detail on the molecular level, especially in carrot and alfalfa suspension cultures (De Jong et al., 1993; Dudits et al., 1991). Somatic embryogenesis from these suspension cultures has been widely applied to study the transition of proembryogenic mass to bipolar embryo that occurs when plant growth hormones are removed from the culture medium (Dudits et al., 1991). However, these cultures are less well suited for studying the earliest events that lead to the induction of embryogenesis, as redetermination of the somatic cells towards embryogenic development and the formation of proembryogenic masses are not synchronous, and already take place while the cells are being grown in the presence of plant growth hormones (Komamine et al., 1990; McWilliam et al., 1974). Direct regeneration of embryos from single somatic cells or protoplasts is possible (Song et al., 1990), but not at high enough efficiency for molecular studies. These problems can be circumvented by studying direct regeneration of embryos from microspores or immature pollen. In microspore culture of oilseed rape (*Brassica napus*), a short heat treatment causes an irreversible, (semi)quantitative and synchronous redetermination of developmental fate from normal pollen development to embryogenic development (Custers et al., 1994; Pechan and Keller, 1988). In tobacco, (*Nicotiana tabacum*) a combination of heat stress and starvation is the most efficient induction condition for microspore embryogenesis (Touraev et al., 1996a). Here, we will try to convince the reader that microspore culture of *B. napus* and *N. tabacum* can serve as excellent model systems for studying the redetermination to embryogenic development in plants. Knowledge of this process will not only be useful in the development of microspore culture protocols for recalcitrant crops and cultivars, but will hopefully also lead to a better understanding of zygotic embryogenesis, and other agronomically important processes such as apomixis, parthenogenesis, and development of artificial seeds.

5.2 Microspore culture

Since the first observation of microspore embryos in anther culture of Datura innoxia in 1964 (Guha and Maheshwari, 1964), the use of anther and microspore culture has become a fixture in the breeding programs of a variety of crops. Microspore derived doubled haploid plants facilitate the rapid isolation of homozygous lines and selection for recessive traits, which reduces cultivar development time (Evans et al., 1990; Hu and Huang, 1987; Huang and Keller, 1989; Morrison et al., 1991; Pickering and Devaux, 1992). Because of the economic importance of doubled haploid plants to plant breeders, research on microspore embryogenesis has been predominantly focussed on the development of efficient culture systems for recalcitrant crops and cultivars (for reviews see: Dickinson, 1992; Huang and Keller, 1989; Sunderland, 1979). Despite the heavy investment into the development of such culture protocols, progress has been relatively slow, and few general principles governing the induction of microspore embryogenesis have been identified (Dunwell, 1992). Two model systems have emerged in which the induction of microspore embryogenesis can be studied with molecular biological and biochemical tools: *N. tabacum* and *B. napus*. The bulk of our understanding of the molecular events that accompany the induction of microspore embryogenesis stems from studies on these two systems.

5.3 Microspore culture of *Brassica napus*

Both genetic and physiological factors influence the efficiency of microspore embryogenesis. The effect of genotype and growth conditions of the donor plant on embryo induction in microspore culture of *B. napus* has been extensively reviewed previously (Chuong and Beversdorf, 1985; Chuong et al., 1988; Gland et al., 1988; Takahata et al., 1991; Vyvadilova and Zelenkova, 1992). Most of the biochemical and developmental studies have been conducted with the cultivar Topas, which shows a high frequency of embryo induction of up to 70 % under optimal culture conditions (Pechan and Keller, 1988). There are two consistent physiological factors which determine to a great extent the efficiency of embryo induction: (1) the developmental stage of the microspores at the time of isolation, and (2) the type and strength of the external trigger applied. For a long time it has been assumed that microspore embryogenesis can be induced only in a narrow window of developmental stages (Pechan and Keller, 1988). But more recent data reveal that the range of developmental stages responsive to embryo induction is much wider than has previously been thought and greatly depends on the type and extent of stress applied to the microspores after isolation. In *B. napus*, a mild heat stress at 32°C for at least 8 h is commonly used to induce sporophytic development in late unicellular microspores and early bicellular pollen. These stages are the most responsive to this heat stress, whereas earlier and later developmental stages show little or no response. Very recently it was demonstrated that also mid to late bicellular pollen of oilseed rape can be induced to enter embryogenic development (Binarova et al., 1997). When a more severe heat stress (41°C) was applied during the first 1-2 h of culture, high numbers of embryos could be produced from these older stages of bicellular pollen.

The possibility of replacing the heat stress treatment for induction of embryogenesis with other stress treatments, such as gamma irradiation or ethanol stress, has been reported, but is much less efficient and only very low numbers of embryos are obtained (Pechan and Keller, 1989). Colchicine is the only inducing agent known to date that induces embryogenesis in microspores of cv. Topas at comparable rates to heat stress induction (Iqbal et al., 1994; Zhao et al., 1996). When using colchicine, the mid unicellular stage was shown to be most responsive (Zhao et al., 1996). By depolymerization of the microtubules this drug disrupts the cytoskeleton of the microspore, resulting in a symmetric division instead of the asymmetric first pollen mitosis which normally occurs during pollen development. Since the colchicine treatment was given at a culture temperature of 25°C in relatively young material, and we have shown that a significant proportion of the microspores divides sporophytically when cultured at 25°C (Custers et al., 1994), the conclusion that colchicine alone can induce embryogenic development in microspores might be somewhat premature.

In conclusion, microspores and pollen of a broad range of developmental stages are responsive to embryo induction. There seems to be an increasing requirement for stress during the induction stage as the microspores mature from the mid unicellular stage to the late bicellular pollen stage. This last observation might be caused by the reduced ability of pollen to mount a heat shock response with increasing age (Mascarenhas and Crone, 1996).

5.4 Microspore culture of *Nicotiana tabacum*

Whereas in microspore culture of oilseed rape temperature stress is the primary inductive trigger, initial experiments with tobacco (Kyo and Harada, 1986) revealed that starvation is the most efficient method for inducing embryogenic development in pollen of this species. Isolated mid bicellular pollen must be cultured for at least three days in a medium without sugar and nitrogen and then transferred to a sugar- and nitrogen containing medium, after which the pollen start to divide sporophytically. Since the composition of the medium determines to a great extent the developmental fate of tobacco pollen in culture, the influence of the various medium components, particularly the carbon and nitrogen source, has been studied extensively (Benito Moreno et al., 1988; Kyo and Harada, 1986; Kyo and Harada, 1990). Benito Moreno et. al. (1988) showed that sucrose is the main determining component of the culture medium, with embryo induction occurring under conditions of sucrose starvation. Additional nitrogen starvation is not required, but it reduces the time necessary for induction of embryogenic pollen.

For a long time it was thought that only tobacco pollen at the mid bicellular stage could be induced to enter embryogenic development. However, one brief report suggested, that also (mid to late) microspores could be induced to enter embryogenic development by giving a prolonged starvation treatment at an elevated culture temperature of 37°C (Vicente et al., 1991). This starvation/heat shock treatment resulted in many symmetrical divisions of the microspores already before their transfer to a medium rich in sucrose. Recently, the induction of embryogenesis in tobacco microspores was studied in more detail (Touraev et al., 1996a). These authors found that microspores of tobacco, like those of *B. napus*, can be induced to enter embryogenic development in rich medium by heat shock alone (33°C or 37°C). However, a combination of starvation and high temperature treatment produced the highest rate of embryogenesis (Touraev et al., 1996a). These authors generally state that tobacco microspore embryogenesis is induced mainly by starvation, and that high temperature has only a minor effect (Touraev et al., 1996a; Touraev et al., 1996b). In our opinion the reverse could be argued equally well, thereby unifying the tobacco and *B. napus* model systems for microspore embryogenesis. We tested in *B. napus* whether high temperature could be replaced or enhanced by starvation as a trigger for embryo induction. Results of these experiments showed that isolated microspores of *B. napus* are not viable in a culture medium with mannitol replacing sucrose. It is therefore still not clear whether starvation can induce embryogenesis from oilseed rape microspores. However, it can now be concluded that the induction of microspore embryogenesis in *N. tabacum* and *B. napus* are brought about by the same basic mechanism, and that the stress treatment has to be adapted to the competence of the starting material.

In our opinion, this analogy between the tobacco and oilseed rape microspore embryogenesis systems may be extended to somatic embryogenesis, and perhaps even to zygotic embryogenesis. It has been shown that embryogenic development can be induced on wounded leaves of *Cichorium* by culturing at a high temperature of 35°C (Decout et al., 1994). When cultured at 20°C or 25°C, these wounded leaves form callus and show shoot organogenesis instead of direct somatic embryogenesis. Temperature of 30°C appeared to be intermediate, showing all types of morphogenesis. Based on the observation of temperature stress induced direct somatic embryogenesis in *Cichorium*, we think that the mechanism by which embryogenesis is induced is the same in both microspores and somatic tissues.

5.5 Molecular events during the induction phase

5.5.1 Changes in gene expression

The induction of embryogenesis in microspores consists of the termination of the existing gametophytic gene expression pattern, and its replacement with a sporophytic expression program. The establishment of two efficient culture systems for microspore embryogenesis, as described above, has opened the possibility to study the mechanisms underlying the developmental switch from gametophytic towards sporophytic development at the molecular level. Over the past few years, a number of biochemical and molecular markers for microspore embryogenesis have been identified (for review, see Cordewener et al., 1996; Reynolds, 1997).

The meaningful analysis of gene expression during embryo induction in microspores is only possible when a comparison can be made with a non-embryogenic control culture. For microspore embryogenesis a good control would be a culture condition in which microspores continue to develop gametophytically. In *B. napus* the culture temperature alone determines the developmental fate of the isolated microspores: culture at 32°C yields embryos and culture at 18°C in the same medium yields mature pollen (control). In tobacco the situation is somewhat more complex since gametophytic and sporophytic development take place in different media, and for embryogenesis the medium has to be changed after the initial induction phase. In addition, the first sporophytic division occurs only after about 10 days in rich medium. This has the advantage, however, that the redetermination phase can be analyzed separately from the subsequent embryogenic phase. Garrido et. al. (1993) did not find any differences in 2-D protein pattern between tobacco pollen labeled with [^{35}S]-methionine at the onset of culture and at the end of a 7 days starvation period. Harada et. al. (1988) analyzed the 2-D protein patterns of tobacco pollen cultured either under starvation or under maturation (control) conditions, but they were also not able to detect any protein spot specific to redifferentiating pollen. One reason for this might be that in these experiments a relatively long period of starvation was needed to acquire embryogenic competence of the pollen, possibly because the starch present in the bicellular pollen first has to be consumed (Sangwan and Sangwan-Norreel, 1987). Another possible reason why no changes in protein pattern have been found during the starvation treatment could be that suppression of protein synthesis is a necessary condition for the redifferentiation of tobacco pollen. Indications for this were given by Harada et. al. (1988) who noted quantitative differences in total protein content and the rate of protein synthesis: during maturation both of them increased, while during starvation both of them decreased. A recent report has shown that tobacco microspores can be induced to enter embryogenic development already within one day of culture by applying a combined starvation/heat shock treatment (Touraev et al., 1996a). This may prove to be a more effective system for studying changes accompanying the induction of embryogenesis at the molecular level in tobacco.

The decrease in overall protein synthesis as described above for the classical method using starved pollen of tobacco is opposite to what has been found with the *B. napus* system. This may be because in *B. napus* the induction of embryogenic development in microspores is far more rapid than in *N. tabacum* pollen culture; the first sporophytic division taking place within the first day in culture in *B. napus* and after about 10 days in *N. tabacum*. A two-fold increase in overall protein synthesis was observed when microspores of *B. napus* were cultured for 8 h under embryogenic conditions (32°C) compared to non-embryogenic conditions (18°C) (Cordewener et al., 1994). Furthermore, a comprehensive analysis of the 2-D protein patterns of both cultures revealed six protein spots which were exclusively synthesized under embryogenic conditions. Another 18 proteins were synthesized at a significantly higher rate under embryogenic conditions (32°C) than the control (18°C). A number of these highly induced proteins were identified as members of the 70 kDa and the low molecular weight class of heat shock proteins (Cordewener et al., 1995; Custers et al., 1996).

Because it has proven difficult to microsequence the differentially expressed protein spots on the 2-D protein gel, a number of groups (including ourselves) are now identifying differentially expressed mRNAs and cloning the corresponding cDNAs that are specific for the induction phase of embryogenesis. Another approach to identify developmental markers for microspore embryogenesis is to make use of probes from genes that are known to be expressed during zygotic embryogenesis. In this way it was shown that expression of a distinct subfamily of napin genes coincides with the induction of *B. napus* microspore embryogenesis (Boutilier et al., 1994). Although it is not clear whether napin gene expression plays a functional role in the induction of microspore embryogenesis, at present it represents the earliest known molecular marker for this process in oilseed rape.

5.5.2 Embryo induction and heat shock proteins

Changes in protein synthesis observed during the induction phase of microspore embryogenesis in *B. napus* are mainly due to an increased synthesis of HSPs (Cordewener et al., 1995). To investigate in more detail the possible role of the 70 kDa class of heat shock proteins in embryo induction, immunolocalization studies with anti-HSP68 and anti-HSP70 were performed on sectioned microspores to determine the subcellular distribution of these HSPs (Cordewener et al., 1995). HSP68, which co-distributed with DNA-containing organelles in the cytoplasm, probably mitochondria, did not show major changes during *in vivo* microspore development. When placed in culture for 8 h at either 18°C or 32°C, comparable HSP68 staining patterns were found. HSP70, on the other hand, showed a developmental-stage-specific intracellular localization, and even more interestingly, its subcellular localization was dependent on the culture conditions. Whereas at non-embryogenic culture conditions (18°C) the distribution of HSP70 was comparable to that found during *in vivo* pollen development, major differences in both the staining intensity and the subcellular localization of HSP70 were observed during temperature-induced (32°C) embryogenesis. During normal pollen development HSP70 was localized in the nucleoplasm during the synthetic (S) phase of the cell cycle, and predominately in the cytoplasm during other phases (Fig. 5.1A). Induction of embryogenesis in late unicellular microspores resulted in an increased overall labeling intensity, and a prolonged anti-HSP70 labeling of the nucleoplasm (Fig. 5.1B). This may be related to an elongation of the S phase of the cell cycle, which is frequently observed after heat shock (Roti Roti et al., 1992; Walsh et al., 1993; Walsh and Morris, 1989). Early bicellular pollen showed a weak generalized cytoplasmic distribution of HSP70 during normal pollen development (Fig. 5.1C), and an accumulation of HSP70 in one (vegetative) or both of the nuclei under embryogenic conditions (Fig. 5.1D). This strong labeling of the nucleoplasm of the vegetative nucleus is probably associated with the re-entry into the cell cycle of the vegetative cell, an event which is required for the onset of embryogenesis (Binarova et al., 1993).

Recently it was shown that pollen isolated at the late bicellular stage can also be induced to enter embryogenic development (Binarova et al., 1997). By the application of an additional short and more severe heat stress at the onset of culture, DNA synthesis is initiated in the vegetative nucleus and followed by sporophytic development and embryo formation. Interestingly,

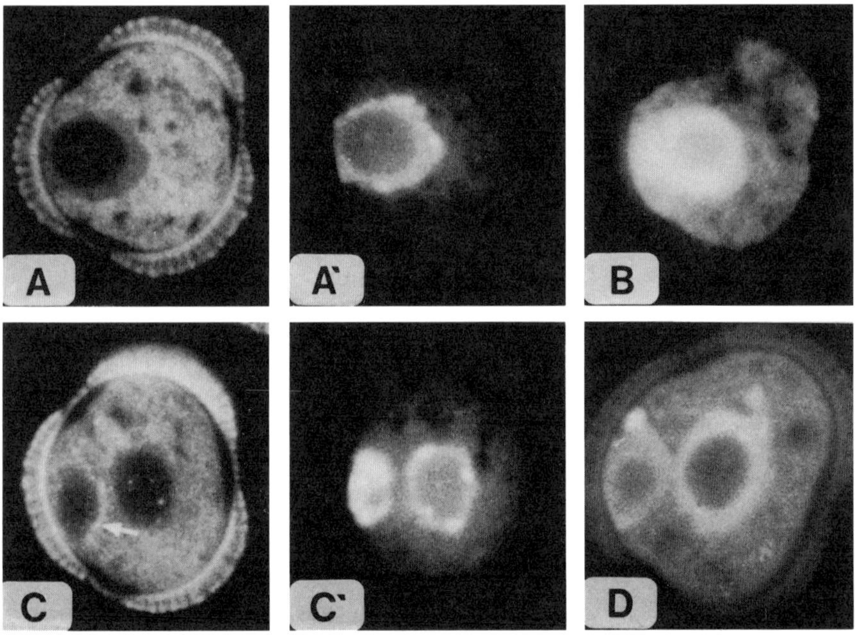

Fig. 5.1. Immunolocalization of HSP70 in late unicellular microspores **A, B** and early bicellular pollen **C, D.**
A-D: HSP70 labeling; **A'**, **C'**: DAPI stained images visualizing the nuclei of photos A and C respectively. **A, A'**: Freshly isolated late unicellular microspore: most of the HSP70 is localized in the cytoplasm. **B**: Late unicellular microspore after 8 h of culture at 32°C: strong HSP70 labeling of the nucleoplasm, and only weak labeling of the cytoplasm. **C, C'**: Freshly isolated early bicellular pollen: HSP70 labeling in the cytoplasm of both the generative and vegetative cell. HSP70 staining was never observed in the vegetative nucleus under these conditions. Note labeled spots near the wall between both cells (arrow). **D**: Early bicellular pollen after 8 h of culture at 32°C: strong HSP70 labeling of both the generative and vegetative nucleus.

immunocytochemistry on cryo-sectioned pollen using anti-HSP70 showed staining of the vegetative nuclei of the late bicellular pollen that had received the additional heat stress (41°C), but not of nuclei of those pollen cultured continuously at 32°C. All these results corroborate the tight correlation between the induction of embryogenesis in microspores or pollen of *B. napus*, and an increased synthesis of HSP70 followed by its translocation to the nucleus during S phase.

Zarsky et al. (1995) studied the activation of a low-molecular weight heat shock gene during embryo induction of tobacco pollen. After isolation of the tobacco clone from a mid bicellular pollen cDNA library, they investigated the expression pattern during normal pollen development and after induction of embryogenesis by starvation. Northern blot analysis showed a positive signal during the final stages of pollen maturation. A more dramatic increase in the transcript was found after embryo induction in mid bicellular pollen (7 days of starvation at 25°C). The level was similar to that in mid bicellular pollen heat shocked for 2 h at 37°C. These results suggest that low-molecular weight HSPs might be associated with the redetermination phase which precedes embryogenesis in tobacco pollen.

Thus, both in the tobacco and oilseed rape culture systems a tight correlation has been found between the induction of embryogenesis and the expression of heat shock genes. It is now necessary to determine whether there is a causal relationship between these observations. It could be that the induction of heat shock gene expression is only a consequence of the stress treatment, and has nothing to do with the induction of embryogenesis. In this scenario, embryogenesis could be induced by the destruction or modification of a factor that is responsible for maintaining pollen development, embryogenic development being the default. The other possibility is that the heat shock gene expression is instrumental in the induction of embryogenesis. In this scenario heat shock gene expression may be required for the reactivation of the cell cycle and the reprogramming of gene expression towards the new developmental pathway.

5.6 The role of heat shock proteins

Plants as well as all other organisms possess HSP genes whose expression is upregulated by environmental stresses such as heat shock or chemicals. There is currently a large body of data available about the general response of cells to stress, and a large array of HSPs have been described. Apart from the heat-inducible HSP genes there are the so-called heat shock cognate (HSC) genes which are constitutively expressed, but show high homology to the inducible HSPs. Since both classes of genes seem to display similar general functions, i.e. in chaperoning proteins during synthesis, (un)folding, assembly and degradation, the term HSP will be used here for both classes of heat shock genes. In this section, experimental data will be presented exemplifying that, apart from their role as molecular chaperones, more distinct functions can be assigned to certain classes of HSPs.

5.6.1 During steroid receptor signalling

It is not always obvious whether HSPs are expressed as a direct result of the stress treatment, or as a response to the subsequent developmental change. For instance in lower eukaryotes like yeast an adverse environmental condition can lead to a new developmental program that leads to the formation of spores. This transition from vegetative growth to differentiation is accompanied by the accumulation of HSP70 (Kurtz et al., 1986). Also, the same treatments as normally used to trigger a heat shock response can initiate sporulation, i.e. heat, starvation, ethanol. Interestingly, similar stress treatments are also effective in switching the developmental fate of microspores from pollen development to embryogenesis. In addition, this change in developmental pathway of the microspores is associated with an increased synthesis of certain classes of HSPs, as discussed above (Cordewener et al., 1995; Zarsky et al., 1995). All this suggests that there is a common pathway between on the one hand the developmental induction of HSPs and on the other hand, their induction by a stress treatment.

Differentiation processes are often controlled by hormones, which can activate or repress the expression of genes via binding to steroid receptors such as the glucocorticoid receptor. Unactivated steroid receptors exist in large heteromeric complexes with heat shock proteins (HSP70 and HSP90), immunophilins and a recently characterized chaperone protein called p23 (Freeman et al., 1996). HSP70 is required for the binding of HSP90 to the steroid receptor and for simultaneous conversion of the unactive receptor to the steroid binding conformation. Binding of hormone stimulates receptor activation, dissociation of HSP90 from the receptor complex, and nuclear translocation of the liganded receptor. Recently, it has been shown that heat shock promotes nuclear accumulation of the hormone-free glucocorticoid receptor (Liu et al., 1995; Sanchez, 1992). Czar et al. (1996) showed in rat cells that HSP90 is able, either directly or indirectly via other proteins, to associate with both microtubules and microfilaments. And in hamster cells it has been shown that application of a heat shock leads to the artificial release of the glucocorticoid receptor from the cytoskeleton, which then allows the (partial) activation of glucocorticoid regulated transcription in the absence of hormone (Shen et al., 1993).

Although there is no evidence that hormones play a role in the signal transduction pathway leading to microspore embryogenesis, it may be that in plants similar mechanisms are operative as the steroid receptor signalling pathway in mammalian systems, and that there is a role for HSPs in the activation of transcription factors. One of the first obvious effects of heat shock on cultured microspores is a disintegration of the cytoskeleton (Hause et al., 1993). A way by which stress could bring about changes in gene expression is through modulation of the localization of transcription factors, for example by causing the disassembly of a glucocorticoid-like heterocomplex bound to the cytoskeleton. HSPs could function here also as chaperones of nuclear transport, facilitating the movement of other proteins across the nuclear membrane. We have shown the translocation of HSP70 to the nucleus of microspores cultured under heat shock conditions (Cordewener et al., 1995). This might be associated with the activation and subsequent nuclear translocation of transcription factor(s) necessary for the initiation of the embryogenic pathway.

5.6.2 During embryogenesis

In higher eukaryotes, fertilization of the egg cell initiates embryogenic development. Studies with mammalian systems showed that transcription is not turned on immediately after fertilization, although some protein synthesis is already observed due to the translation of maternal mRNAs. In mouse, the onset of zygotic gene expression occurs on the second day after fertilization, at the two-cell stage during which a specific set of proteins is synthesized. Two of these proteins correspond to the inducible HSP70 and the cognate HSC70 (Bensaude et al., 1983). One day later, at the eight-cell embryo stage, the synthesis of the inducible HSP70 has become undetectable even after heat shock, whereas the cognate HSC70 is still synthesized at a very high level. But even within one cell cycle changes in the expression of HSPs have been detected, both at the mRNA and protein level. Milarski and Morimoto (1986) showed in synchronized human cells a rapid increase in the level of HSP70 upon entry into S phase and a decline by late S and G2. Indirect immunofluorescence with HSP70-specific monoclonal antibodies revealed that the protein accumulated in the nucleus of early S phase cells, whereas it was diffusely distributed throughout the cell during the remainder of the cell cycle.

Cell cycle specific changes in the subcellular distribution of HSP70 were also observed during pollen development in *B. napus* (Cordewener et al., 1995). Indirect immunofluorescence using anti-HSP70 revealed a strong labeling of the nucleus compared to the cytoplasm of vacuolate late unicellular microspores. During the process of vacuole destruction, just before the first pollen mitosis, HSP70 labeling in the cytoplasm increased while it gradually decreased in the nucleus. During mitosis, HSP70 was equally distributed in the cytoplasm. Early bicellular pollen showed an intensive fluorescence of the cytoplasm, but no labeling of the generative and vegetative nuclei.

After heat shock treatment of mammalian cells, HSP70 accumulates within the nucleus and specifically associates with nucleoli (Milarski and Morimoto, 1986). This dramatic redistribution of HSP70 from the cytoplasm to the nucleus occurs independently of the phase of the cell cycle at which the heat shock is applied. Induction of embryogenesis in microspores and pollen of *B. napus* by heat shock is also accompanied by the accumulation of HSP70 within the nucleus (Cordewener et al., 1995), but contrary to the observations in mammalian cells the protein is only found in the nucleoplasm. Thus, unlike in the heat shock response of mammalian cells, in which HSP70 is found tightly bound to nucleolar structures following nuclear migration, HSP70 does not seem to interact strongly with nucleolar structures during embryo induction in microspores. It is worth noting that this very prominent nucleolar deposition of HSP70 in mammalian cells is only detected after heat shock, and a more generalized distribution of HSP70 throughout the nucleus is found in S-phase cells cultured at physiological temperature (Milarski and Morimoto, 1986; Milarski et al., 1989). The heat treatment used in *B. napus* microspore culture for triggering embryogenesis is not a 'real' heat shock since no sublethal temperature is applied to the cells. This is in conformity with the finding that the overall protein synthesis during the first 8 h of culture at 32°C increases twofold as compared to a 18°C culture (Cordewener et al., 1994), an observation that contrasts with the typical heat shock response in which the induction of heat shock protein synthesis is accompanied by a suppression of the synthesis of most other cellular proteins. All these data suggest that the function of HSP70 in the nucleus of embryo-induced microspores or pollen (after a mild heat stress) may be distinct from that of cells subjected to a sublethal heat treatment.

While most plant HSPs are similar to HSPs found in other organisms, an unusual feature of plants is the presence of numerous low molecular weight HSPs. Accumulation of HSP18 during the starvation treatment of tobacco pollen suggests a possible role of low molecular weight HSPs in the redifferentiation process preceding embryo formation. Expression and localization studies at the protein level are required in order to determine the function of this class of HSPs during embryo induction in tobacco.

5.7 Conclusions and perspectives

In conclusion, stress seems to be the common factor in triggering the sporophytic pathway in microspores from various sources, including oilseed rape and tobacco. The strong similarity of the stress treatments necessary for the induction of embryogenesis in microspores of the two last species suggests a common molecular pathway. Until now research on the role of stress proteins during embryo induction in *B. napus* microspores has focussed on the 70 kDa class of HSPs, whereas for the tobacco system only data are available about the low molecular weight HSPs. Both classes of HSPs might be involved in the induction process. Knowledge about the different classes of HSPs from other systems is accumulating, especially about the protective role that HSP70 fulfills in mammalian cells (Feige and Plla, 1994). It is obvious now that HSPs are essential for the survival of cells and organisms under normal as well as under stressful conditions. A mild stress can prepare a cell for coping better with a more severe stress. In microspore culture of both oilseed rape and tobacco a mild stress blocks the gametophytic pathway and induces sporophytic development. Although much more research has to be done to unravel the underlying molecular mechanisms, sufficient knowledge about both model systems is currently available to start devising common strategies. Including in such research the molecular analysis of somatic embryogenesis induction in *Cichorium* will likely reveal that HSPs have a general function in the developmental switch which leads to embryogenesis.

Acknowledgements

This work was supported by grant PL960275 from the European Union Framework IV Biotechnology Programme.

References

Benito Moreno RM, Macke F, Hauser MT, Alwen A, Heberle-Bors E (1988) Sporophytes and male gametophytes from in vitro cultured, immature tobacco pollen. In: Sexual reproduction in higher plants. Springer-Verlag, New York pp 137 – 142

Bensaude O, Babinet C, Morange M, Jacob F (1983) Heat shock proteins, first major products of zygotic gene activity in mouse embryo. Nature 305:331 – 333

Binarova P, Hause G, Cenklova V, Cordewener JHG, Van Lookeren Campagne MN (1997) A short severe heat shock is required to induce embryogenesis in late bicellular pollen of Brassica napus L. Sex Plant Reprod 10:200 – 208

Binarova P, Straatman K, Hause B, Hause G, Van Lammeren AAM (1993) Nuclear DNA synthesis during the induction of embryogenesis in cultured microspores and pollen of Brassica napus L. Theor Appl Genet 87:9 – 16

Boutilier KA, Gines MJ, Demoor JM, Huang B, Baszczynski CL, Iyer VN, Miki BL (1994) Expression of the BnmNAP subfamily of napin genes coincides with the induction of Brassica microspore embryogenesis. Plant Mol Biol 26:1711 – 1723

Chuong PV Beversdorf WD (1985) High frequency embryogenesis through isolated microspore culture in Brassica napus L. and B. carinata Braun. Plant Sci 39: 219 – 226

Chuong PV, Deslauriers C, Kott LS, Beversdorf WD (1988) Effects of genotype and bud sampling on microspore culture of Brassica napus. Can J Bot 66:1653 – 1657

Clark JK, Sheridan WF (1991) Isolation and characterization of 51 embryo-specific mutations of maize. Plant Cell 3:935 – 951

Cordewener JHG, Busink R, Traas JA, Custers JBm, Dons JJM, Van Lookeren Campagne MM (1994) Induction of microspore embryogenesis in Brassica napus is accompanied by specific changes in protein synthesis. Planta 195:50 – 56

Cordewener JHG, Custers JBM, Dons HJM, Van Lookeren Campagne MM (1996) Molecular an biochemical events during the induction of microspore embryogenesis. In: In vitro haploid production in higher plants, Kluwer Academic Publishers, Dordrecht pp111 – 124

Cordewener JHG, Hause G, Görgen E, Busink R, Hause B, Dons HJM, Van Lammeren AAM, Van Lookeren Campagne MM, Pechan P (1995) Changes in synthesis and localization of the 70 kDa class of heat shock proteins accompany the induction of embryogenesis in Brassica napus L. microspores. Planta 196:747 – 755

Custers JBM, Cordewener HG, Dons HJM, Van Lookeren Campagne MM (1996) Regulation of the inductive phase of microspore embryogenesis in Brassica napus. Acta Hort 407:209 – 217

Custers JBM, Cordewener JHG, Nöllen Y, Dons JJM, Van Lookeren Campagne MM (1994) Temperature controls both gametophytic and sporophytic development in microspore cultures of Brassica napus. Plant Cell Rep 13:267 – 271

Czar MJ, Welsh MJ, Pratt WB (1996) Immunofluorescence localization of the 90 kDa heat shock protein to cytoskeleton. Eur J Cell Biol 70:322 – 330

De Jong AJ, Schmidt DL, De Vries SC (1993) Early events in higher-plant embryogenesis. Plant Mol Biol 22:367 – 377

Decout E, Dubois T, Guedira M, Dubois J, Audran JC, Vasseur J (1994) Role of temperature as a triggering signal for organogenesis or somatic embryogenesis in wounded leaves of chicory cultured in vitro. J Exp Bot 45:1859 – 1865

Dickinson HG (1992) Microspore derived embryogenesis. In: Sexual Plant Reproduction, Springer-Verlag, Berlin pp1 – 15

Dudits D, Bögre L, Györgyey J (1991) Molecular and cellular approaches to the analysis of plant embryo development from somatic cells in vitro. J Cell Sci 99:475 – 484

Dunwell JM (1992) Mechanisms of microspore embryogenesis. In: Reproductive biology and plant breeding, Springer-Verlag, Berlin pp121 – 130

Evans DE, Singh MB, Knox RB (1990) Pollen development: applications in biotechnology. In: Microspores. Evolution and ontogeny, Academic Press, San Diego pp309 – 338

Faure JE, Digonnet C, Dumas C (1994) An in vitro system for adhesion and fusion of maize gametes. Science 263:1598 – 1600

Feige U, Plla BS (1994) Heat shock proteins: the hsp70 family. Hsp70 - a multi-gene, multi-structure, multi-function family with potential clinical applications. Experientia 50:979 – 985

Freeman BC, Toft DO, Morimoto RI (1996) Molecular chaperone machines: chaperone activities of the cyclophilin Cyp-40 and the steroid apo-receptor-associated protein p23. Science 274:1718 – 1720

Garrido D, Eller N, Heberle-Bors E, Vicente O (1993) De novo transcription of specific messenger RNAs during the induction of tobacco pollen embryogenesis. Sex Plant Reprod 6:40 – 45

Gland A, Lichter R, Schweiger HG (1988) Genetic and exogenous factors affecting embryogenesis in isolated microspore cultures of Brassica napus L. J Plant Physiol 132:613 – 617

Guha S, Maheshwari SC (1964) In vitro production of embryos from anthers of Datura. Nature 204:497

Harada H, Kyo M, Imamura J (1988) The induction of embryogenesis in Nicotiana immature pollen culture. In: Applications of plant cell and tissue culture, Wiley, Chichester pp59 – 74

Hause B, Hause G, Pechan P, Van Lammeren AAM (1993) Cytoskeletal changes and induction of embryogenesis in microspore and pollen cultures of Brassica napus L. Cell Biol Int 17:153 – 168

Hu H, Huang B (1987) Application of pollen derived plants to crop improvement. Int Rev Cytol 107:293 – 313

Huang B, Keller WA (1989) Microspore culture technology. J. Tissue Cult Meth 12:171 – 178

Iqbal MCM, Möllers C, Robbelen G (1994) Increased embryogenesis after colchicine treatment of microspore cultures of Brassica napus L. J Plant Physiol 143:222 – 226

Jürgens G. Mayer U, Torrez Ruiz RA, Berleth T, Miséra S0 (1991) Genetic analysis of pattern formation in the Arabidopsis embryo. Development Suppl 1:27 – 38

Komamine A, Matsumoto M, Tsukahara M, Fujiwara A, Kawahara R, Ito M, Smith J, Nomura K, Fujimura T (1990) Mechanisms of somatic embryogenesis in cell cultures - physiology, biochemistry and molecular biology. In: Progress in Plant Cellular and Molecular Biology, Kluwer Acad. Publ., Dordrecht pp307 – 313

Kranz E, Lorz H (1993) In vitro fertilization with isolated, single gametes results in zygotic embryogenesis and fertile maize plants. Plant Cell 5:739 – 746

Kurtz S, Rossi J, Petko L, Lindquist S (1986) An ancient developmental induction in Saccharomyces sporulation and Drosophila oogenesis. Science 231:1154 – 1157

Kyo M, Harada H (1986) Control of the developmental pathway of tobacco pollen in vitro. Planta 168:427 – 432

Kyo M, Harada H (1990) Specific phosphoproteins in the initial period of tobacco pollen embryogenesis. Planta 182:58 – 63

Laux T, Jürgens G (1994) Establishing the body plan of the Arabidopsis embryo. Acta Bot. Neerl 43:247 – 260

Liu W, Hillmann AG, Harmon JM (1995) Hormone-independent repression of AP-1-inducible collagenase promoter activity by glucocorticoid receptors. Mol Cell Biol 15:1005 – 1013

Mascarenhas JP, Crone DE (1996) Pollen and the heat shock response. Sex Plant Reprod 9:370 – 374

McWilliam AA, Smith SM, Street HE (1974) The origin and development of embryoids in suspension cultures of carrot (Daucus carota). Ann Bot 38:243 – 250

Meinke DW (1991) Perspectives on genetic analysis of plant embryogenesis. Plant Cell 3:857 – 866

Milarski KL, Morimoto RI (1986) Expression of human HSP70 during the synthetic phase of the cell cycle. Proc Natl Acad Sci USA 83:9517 – 9521

Milarski KL, Welch WJ, Morimoto RI (1989) Cell cycle-dependent association of HSP70 with specific cellular proteins. J Cell Biol 108:413 – 423

Morrison RA, Evans DA, Fan ZG (1991) Haploid plants from tissue culture: application in crop improvement. In: Sub-cellular Biochemistry, Plenum Press, New York pp53 – 72

Pechan PM, Keller WA (1988) Identification of potentially embryogenic microspores in Brassica napus. Physiol Plant. 74:377 – 384

Pechan PM, Keller WA (1989) Induction of microspore embryogenesis in Brassica napus L. by gamma irradiation and ethanol stress. In Vitro Cellul Dev Biol 25:1073 – 1074

Pickering RA, Devaux P (1992) Haploid production: approaches and use in plant breeding. In: Barley: Genetics, Biochemistry, Molecular Biology and Biotechnology, C.A.B. International, Oxon pp519 – 547

Reynolds TL (1997) Pollen embryogenesis. Plant Mol Biol 33:1 – 10

Roti Roti JL, Mackey MA, Higashikubo R. (1992) The effects of heat shock on cell proliferation. Cell Prolif 25:89 – 99

Sanchez ER (1992) Heat shock induces translocation to the nucleus of the unliganded glucocorticoid receptor. J Biol Chem 267:17 – 20

Sangwan RS, Sangwan-Norreel BS (1987) Ultrastructural cytology of plastids in pollen grains of certain androgenic and non-androgenic plants. Protoplasma 138:11 – 22

Shen P, Xie ZJ, Li H, Sanchez ER (1993) Glucocorticoid receptor conversion to high affinity nuclear binding and transcription enhancement activity in chinese hamster ovary cells subjected to heat and chemical stress. J Steroid Biochem Mol Biol 47:55 – 64

Song J, Sorensen EL, Liang GH (1990) Direct embryogenesis from single mesophyll protoplasts in alfalfa (Medicago sativa L.). Plant Cell Rep 9:21 – 25

Sunderland N (1979) Comparative studies of anther and pollen culture. In: Plant cell and tissue culture: principles and applications, Ohio State Universit, Ohio pp203 – 219

Takahata Y, Brown DCW, Keller WA (1991) Effect of donor plant age and inflorescence age on microspore culture of Brassica napus L. Euphytica 58:51 – 55

Touraev A, Ilham A, Vicente O, Heberle-Bors E (1996a) Stress-induced microspore embryogenesis in tobacco: an optimized system for molecular studies. Plant Cell Rep 15:561 – 565

Touraev A, Pfosser M, Vicente O, Heberle-Bors E (1996b) Stress as the major signal controlling the developmental fate of tobacco microspores: towards a unified model of induction of microspore/pollen embryogenesis. Planta 200:144 – 152

Vicente O, Benito Moreno RM, Heberle-Bors E (1991) Pollen cultures as a tool to study plant development. Cell Biol Rev 25:295 – 305

Vyvadilova M, Zelenkova S (1992) Responsiveness in microspore cultures of some cultivars and Czech breeding materials of rapeseed (Brassica napus L.). Genet a Slecht. 28:243 – 252

Walsh D, Li K, Wass J, Dolnikov A, Zeng F, Li Z, Edwards M (1993) Heat-shock gene expression and cell cycle changes during mammalian embryonic development. Dev Genet. 14:127 – 136

Walsh DA, Morris VB (1989) Heat shock affects cell cycling in the neural plate of cultured rat embryos: a flow cytometric study. Teratology 40:583 – 592

Zarsky V, Garrido D, Eller N, Tupy J, Vicente O, Schöffl F, Heberle-Bors E (1995) The expression of a small heat shock gene is activated during induction of tobacco pollen embryogenesis by starvation. Plant Cell Environment 18:139 – 147

Zhao JP, Simmonds DH, Newcomb W (1996) Induction of embryogenesis with colchicine instead of heat in microspores of Brassica napus L. cv. Topas Planta 198:433 – 439

6 Mechanisms of Asymmetric Division and Cell Fate Determination in Developing Pollen

David Twell and Ross Howden
Department of Botany, University of Leicester, LE17RH, UK

6.1 Introduction

The generation of cell diversity during development requires that daughter cells of a common division achieve different fates. Asymmetry in the fate of daughter cells may be established by one of two mechanisms: first, by the unequal division of intrinsically polar mother cells, and second, by symmetrical division of non-polar mother cells followed by the imposition of asymmetric external signals (reviewed in Horvitz and Herskowitz, 1992). In plants differences in cell fate arising from the first mechanism are particularly apparent during zygotic division and embryogenesis (Mayer et al., 1991), stomatal complex development (Palevitz, 1982) and during pollen development (Eady et al., 1995). Significant progress has recently been made through the identification of mutants with altered cell fate and the cloning of genes which are required for asymmetric division of the zygote (Shevell et al., 1994) and in the embryogenic and postembryonic root meristem (Di Laurenzio et al., 1996). Our interest is in elucidating the mechanisms establishing cell polarity, asymmetric cell division and cell fate during pollen development. The aims of this chapter are to discuss: the significance of asymmetric cell division for correct pollen differentiation; how asymmetric division leads to differential cell fate; how microspores develop the necessary polarity required for asymmetric division; and to summarize the results of new approaches that are being adopted to identify key genes involved in these processes.

6.2 Pollen cell lineage and development

The completion of meiosis of the diploid microsporocytes within the anther loculus marks the initiation of a unique process of cellular differentiation which ultimately leads through a simple cell lineage to the formation of a large number of male gametophytes shed from the plant as pollen grains (Fig. 6.1A). In the majority of angiosperms this pathway involves only a single asymmetric mitotic cell division, pollen mitosis I (PMI), which serves as a developmental marker for the termination of microspore development and the initiation of pollen development. After pollen germination the generative cell completes a further mitotic division in the pollen tube to form the two sperm cells. In most graminaceous species and certain dicots including *Arabidopsis thaliana* generative cell division occurs precociously prior to pollen shed (Brewbaker, 1967). Since major cytological processes, such as vacuolation, cell wall synthesis, microspore nuclear migration and the accumulation of metabolic reserves are often conserved between species which shed pollen in a bicellular or tricellular condition (Schrauwen et al., 1990; Regan and Moffatt, 1990; Eady et al., 1995; Owen and Makaroff, 1995), it is likely that many gametophytically expressed gene regulatory systems are also conserved. Detailed molecular-genetic analysis of pollen development in the model species *Arabidopsis thaliana* is therefore expected to provide the most efficient route to isolate gametophytic genes involved in asymmetric division at PMI and differential cell fate. The expression and action of such genes then may subsequently be analyzed in other bicellular and tricellular species that are more suitable for cell biological analysis. Details of pollen cell lineage relationships and key differentiation steps in the pathway of pollen development in *Arabidopsis* are illustrated in Fig. 6.1.

Despite the absence of symplastic connections between the sporophytic tissues of the anther and the developing microspores interactions between the sporophytic and gametophytic generations are fundamentally important for the completion of pollen development. For example the tapetum actively secretes and controls the deposition of proteins, lipids, carbohydrates and secondary metabolites, which are required for membrane synthesis, energy requirements and the elaboration of the unique pollen wall (reviewed in Pacini, 1990). The tapetum and anther loculus thus provide essential metabolites and a suitable environment in which differentiation of the microspores can occur. For example, many sporophytically determined mutations act in the tapetum and anther wall and lead to male sterility (Chaudhury, 1993). At later stages of both microspore and pollen development apparently normal differentiation can been achieved in several species in *in vitro* culture if simple essential nutrients are provided (Tanaka and Ito, 1980, 1981; Kyo and Harada, 1986; Benito Moreno et al., 1988; Tupy et al., 1991; Dupuis and Pace, 1993). Thus, the male gametophyte from its inception at PMI, and perhaps to a lesser extent the haploid microspore, is capable of directing its own genetic programme to achieve maturity. This degree of autonomy is reflected in the highly active haploid gene expression programme required for the highly specialized structure and function of the pollen grain. The patterns and diversity of haploid gene expression during male gametophyte development have been comprehensively reviewed, including the functional identification of conserved transcriptional regulatory elements present within the promoter regions of pollen-specific genes (Mascarenhas, 1992; McCormick, 1993; Twell, 1994).

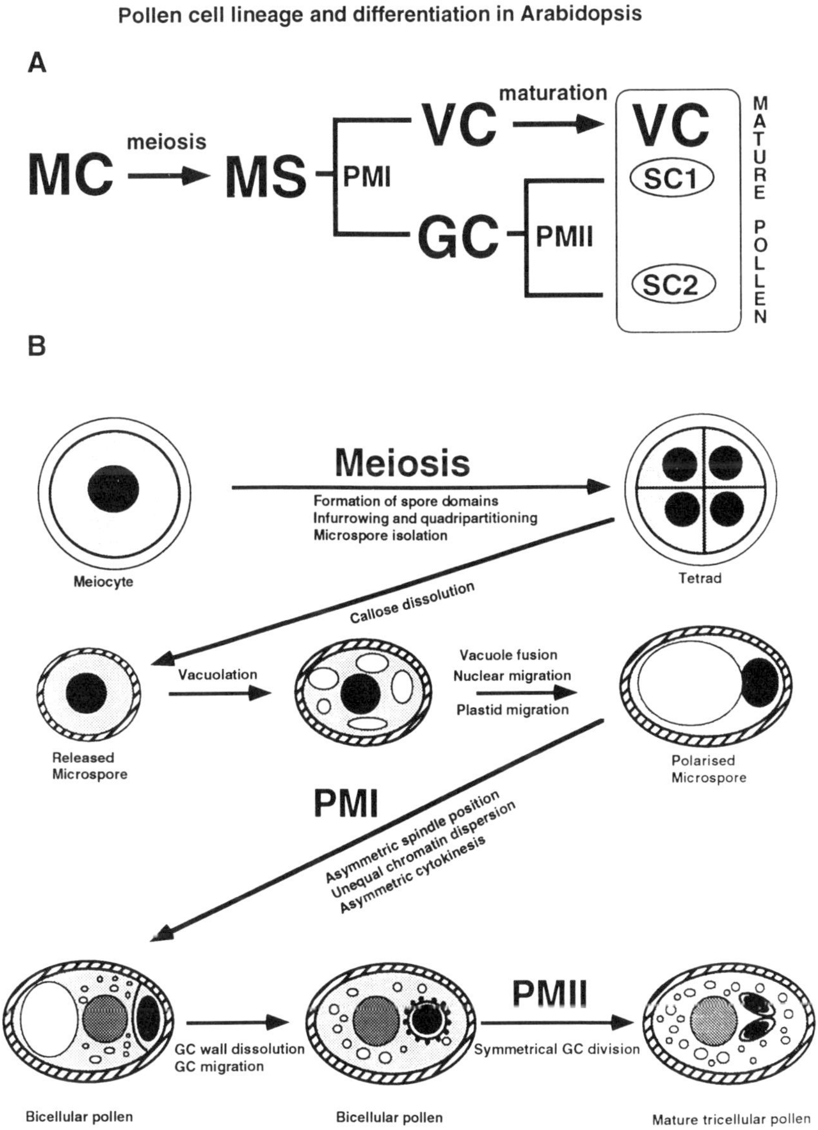

Fig. 6.1. Pollen cell lineage and differentiation. **A** Shows the simple cell lineage relationships in the pathway of pollen development. **B** Diagrammatically illustrates major cytological events which occur during microspore and pollen development in Arabidopsis (Owen and Makaroff, 1995). MC = meiocyte; MS = microspore; VC = vegetative cell; GC = generative cell; SC = sperm cell.

6.3 The asymmetric cell division in pollen

A key event in the differentiation of the angiosperm male gametophyte is the unique asymmetric mitotic division of the haploid microspore termed pollen mitosis I (PMI). This highly asymmetric division produces two unequal daughter cells, the vegetative and generative cells, that have dramatically different structures and developmental fates. The large vegetative cell (VC) constitutes the bulk of the pollen cytoplasm and contains an abundance of stored metabolites required for rapid pollen tube extension. In contrast, the diminutive generative cell (GC) is completely enclosed by the VC cytoplasm and contains relatively few organelles and stored metabolites (Fig. 6.1).

Although considerable information exists concerning the ultrastructural differentiation of the vegetative and GCs as daughter cells of a common mitotic division, details of the mechanisms that determine their respective fates remain largely unknown. The extreme dimorphism in cytoplasmic volume, organelle, protein and RNA content clearly suggests that these cells have very different synthetic and genetic activities. This is supported by ultrastructural analyses, cytochemical RNA staining and the analysis of protein and histone profiles of partially purified vegetative and GCs (reviewed in Mascarenhas, 1975; Sunderland and Huang, 1987).

6.3.1 Differential gene expression in vegetative and generative cells

Direct evidence for the presence of specific transcripts within the vegetative cytoplasm of a number of species has been provided by the *in situ* hybridisation of cDNA clones to developing, mature or germinated pollen (Hanson et al., 1989; Raghavan et al., 1989; Ursin et al., 1989; Brown and Crouch, 1990; Olmedilla et al., 1991; Reijnen et al., 1991; Theerakulpisut et al., 1991; Shen and Hsu, 1992). Although in these studies it was not possible to determine whether transcripts were also present in the generative or sperm cell cytoplasm, it is likely that given their abundance and putative roles in pollen tube growth that these genes are expressed specifically in the VC. Direct evidence for differential gene expression in the vegetative and GCs was obtained by linking the promoter of the tomato *lat52* gene to a GUS fusion protein containing the tobacco etch virus NIa nuclear-targeting signal. The *lat52* promoter was shown to be activated specifically from the vegetative nucleus of tobacco, and not in the GC, neither during pollen development or tube growth (Twell, 1992). The VC-specific activity of the *lat52* promoter was also shown to be maintained in *Arabidopsis* (Eady et al., 1994), and VC-specific activity was similarly demonstrated for the promoter of the tobacco pollen-specific Ntp303 gene (Weterings, 1994).

Several studies have shown that generative and sperm cells actively synthesize RNA and protein at the onset of germination (Mascarenhas, 1966; Reynolds and

Raghavan 1982; Haskell and Rogers, 1985; Zhang et al., 1993). More detailed studies have shown that isolated GCs express unique basic histone-like proteins (Ueda and Tanaka, 1995a,b) and possess their own pool of translatable mRNAs some of which encode proteins unique to the GC (Blomstedt et al., 1996). Recently the first definitive identification of a gene expressed specifically in the GC has been obtained by *in situ* hybridization of a cDNA designated clone #231 isolated from a cDNA library prepared from embryogenic microspores of *Brassica napus* (Havlicky et al., 1996). The full extent to which gene expression profiles of the VC and GC overlap remains obscure, but using a genetic approach several independent mutations specifically affecting generative cell development have been isolated (Section 6.8). This approach therefore provides further access to the repertoire of important genes specifically required for GC and sperm cell development. Taken together these data now provide compelling evidence for the unique gene expression programs of the vegetative and GCs.

6.3.2 The importance of asymmetric division for differential cell fate

Experimental evidence which supports the significant role of the division asymmetry at PMI have come from studies in which symmetrical divisions have been induced by the application of microtubule inhibitors such as colchicine (Tanaka and Ito, 1981; Zaki and Dickinson, 1991), or by physical means such as centrifugation (Terasaka and Niitsu, 1987). The two apparently identical cells resulting from symmetrical division do not possess the nuclear characteristics of the GC but more closely resemble VCs. These data clearly show that division asymmetry is essential for differential cell fate following PMI

Further insight into the differentiated state of daughter cell resulting from unequal and equal division at PMI has been gained through the application of transgenic plants expressing gus activity under the control of the VC-specific *lat52* promoter to monitor cell fate (Twell, 1992; Eady et al., 1995). Such transgenic lines have been used to determine the significance of both the onset and asymmetry of PMI for correct GC differentiation and the activation of VC-specific gene expression.

When division at PMI was blocked in tobacco microspores matured in the presence of the microtubule inhibitor colchicine uninucleate "pollen grains" were produced which were capable of *in vitro* germination (Eady et al., 1995). These results demonstrated first, that a functional GC is not required for pollen maturation, germination or pollen tube growth; and second, since the *lat52-gus* marker was activated in uninucleate "pollen grains", that VC-specific gene expression can be activated independently of division at PMI. This uncoupling of late pollen gene activation from the division process demonstrates that these processes, although normally activated in concert, are subject to independent control. Late gene activation is therefore likely controlled by gametophytically-expressed transcription factors which accumulate in the developing microspore and which normally reach a threshold of activity at PMI by a pathway independent of cytokinesis (see Fig. 6.2).

When symmetrical divisions at PMI are induced with limiting concentrations of colchicine, *lat52-gus* is activated in both daughter cells, and both daughter nuclei show the dispersed "vegetative-like" chromatin appearance (Eady et al., 1995). Therefore symmetrical division produces apparently identical daughter cells which express vegetative cell fate. These data firmly establish that the symmetry of division at PMI plays a vital role in determining GC differentiation, including the repression of VC-specific genes and chromatin dispersion. The extent to which equal vegetative-like cells resulting from induced symmetrical division are in fact equal can be judged since only one cell of the pair is capable of germination (Touraev et al. 1995). This suggests that equal daughter cells are in fact unequal with regard to their maturation with one daughter being incompletely determined, perhaps as a result of the limiting concentrations of gametophytic factor (GF) or the presence of a generative cell repressor (GCR) according to the models described below (Section 6.4.1; Fig. 6.2).

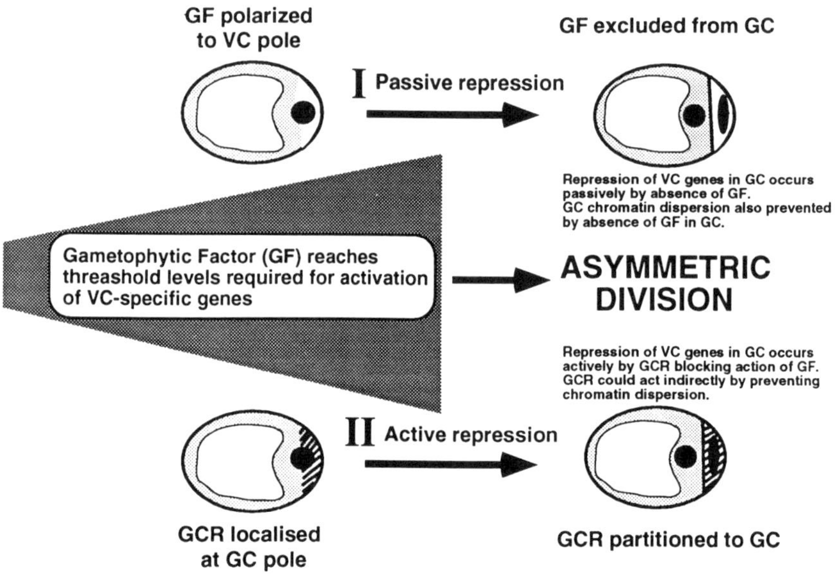

Fig. 6.2. Mechanisms involved in cell fate determination in pollen. In both models, VC-specific gene expression is activated through the accumulation of a gametophytic transcription factor (GF). The repression of VC-specific genes and chromatin condensation in the GC then results either from exclusion of GF from the GC pole (I. Passive Repression Model), or from the asymmetric localization of a generative cell repressor (GCR) protein at the GC pole (II. Active Repression Model).

6.4 Mechanisms of differential cell fate determination

How then does asymmetric division result in the repression of VC gene expression specifically in the GC and the differential dispersion of chromatin within the vegetative and generative nuclei?

6.4.1 Repression of VC gene expression in the GC

Two general models have been proposed to account for the differential activation of VC-specific genes as a result of asymmetric cell division (Eady et al., 1995). Both models are based upon the underlying proposition that the activation of VC genes is the default pathway resulting from the accumulation of a gametophytic transcription factor (GF) to threshold levels at PMI. These models provide alternative mechanisms, both involving the polarized distribution of gametophytic regulatory factor(s) to explain how VC-specific genes are normally repressed in the GC as a result of asymmetric division.

In the passive repression model a gametophytic factor (GF) is excluded from a cytoplasmic domain at the GC-pole (Fig. 6.2). Repression of VC-specific gene activity then results from the absence of GF in the nascent GC. In the active repression model a generative cell repressor (GCR) which is concentrated at the GC-pole and acts to block VC-specific gene activation by GF in the GC (Fig. 6.2). Although these models are necessarily general at this point requiring elaboration with molecular details, they provide a useful framework for designing new experimental approaches and evaluating new evidence in this field.

6.4.2 Differential chromatin dispersion

The above models must also take into account the highly condensed nature of the GC chromatin. The compact GC chromatin structure is likely to limit the access of transcription factors to many genes and may therefore be directly responsible for the maintenance of the transcriptionally repressed condition of the GC. One clue as to the origin of this differential chromatin structure comes from the observation that in over 40 species examined, including gymnosperms, which display multiple asymmetric divisions of the microspore, differential dispersion of the VC and GC chromatin could be observed from late anaphase to late telophase during mitosis (Terasaka and Tanaka, 1974; Terasaka, 1982). Chromatin dispersion therefore appears to be differentially controlled by processes occurring within the same cytoplasm *before* the chromosomes are enclosed within the daughter nuclei. This suggests polarity in the microspore cytoplasm immediately before division. Early evidence for such polarity was provided by the observations of LaCour (1949) who observed that the generative nucleus developed in a region of cytoplasm relatively free of RNA while the cytoplasm surrounding the vegetative nucleus was rich in

RNA. This suggests that during microspore development a "GC domain" is established at the future GC pole.

Within this "GC domain" specific factors must exist which act to limit chromatin dispersion. Unequal chromatin behaviour could therefore involve a cytoplasmic gradient of a GC pole localized "condensation factor". This factor could prevent chromatin dispersion, indirectly via modification of a chromatin component or directly by becoming chromatin associated during mitosis. In the most simple model this "condensation factor" would be identical with the GCR, in which case the GCR would control VC-specific gene expression indirectly by limiting GC chromatin dispersion.

Histone or histone-associated factors are putative candidates for such a "condensation factor", in particular because of the known ability of histones to regulate chromatin packaging. Evidence for differences in histone content between the VC and GC nuclei were reported in studies of the differential cytochemical staining of histones in pollen nuclei of *Lilium* (Jalouzot, 1969) and *Paoenia* (Sauter, 1969). Sheridan (1973) examined histones profiles in isolated VC and GC nuclei of *Lilium* by polyacrylamide gel electrophoresis and amino acid analysis and concluded that the GC nucleus contained a unique lysine-rich histone which was not present in VC nuclei fractions. Recent studies have confirmed and extended these findings in that two histone variants gH2B (lysine rich) and gH3 (arginine rich) purified from GC nuclei of lily have been shown to be present specifically in GC and sperm cell nuclei (Ueda and Tanaka, 1995a, b). These histone variants therefore may be involved in establishing and/or maintaining the altered chromatin structure in GC nuclei. In this regard, interesting parallels can be made with the mechanisms of tight packaging of chromatin in animal sperm, which in sea urchins is correlated with the presence and reversible phosphorylation of two male-germ-line-specific histones (spH1 and spH2B) at tetrapeptide "SPKK" repeat motifs (Poccia and Green, 1992).

6.4.3 Nuclear/cytoplasmic ratio?

A further key difference between the newly formed vegetative and GCs is their remarkably different nuclear:cytoplasmic (N:C) ratios and mechanisms linking N:C ratio to gene expression and chromatin dispersion can be envisioned. For example nuclear chromatin dispersal and VC-specific gene expression may be limited in the GC by the absolute amount of cytoplasmic factors (including transcription factors) rather than their concentration. However, since differential chromatin dispersion can be observed before cytokinesis (Terasaka and Tanaka, 1974) this would suggest that chromatin dispersion is at least initially not controlled directly by N:C ratio. It is perhaps more likely, as discussed above, that a GC domain is established at the future GC pole which through the presence or absence of specific factors limits chromatin dispersion and represses gene expression. However after division the N:C ratio could still play a role in maintaining the condensed state and the repression of VC-specific genes since there is a strong positive correlation between chromatin dispersion and cell size in more or less asymmetric divisions at PMI (Terasaka and Niitsu, 1987). Therefore, purely topological constraints may still contribute to the maintenance of GC chromatin condensation following division.

6.5 Evidence for microspore polarity

Given the strict dependence of generative cell fate on asymmetric division a key question is how does the microspore develops the necessary polarity required to achieve asymmetric division? Microspore polarity is ultimately realized at the completion of asymmetric division. The approach adopted in the following discussion is to trace back from this point to the origins of polarity by systematically identifying cellular components and processes which must be coordinated to achieve asymmetric division.

6.5.1 Spindle asymmetry

The most advanced expression of polarity comes from the observation that the spindle is always asymmetric at PMI being blunt at the GC pole and sharp at the VC pole (Geitler, 1935; Brumfield, 1941; Heslop-Harrison, 1968). Since the cell plate is formed in an asymmetric position closer to the GC pole, spindle asymmetry is therefore likely to be involved in positioning the cell plate through the directed transport and fusion of phragmoplast vesicles. Spindle asymmetry itself may be caused by the proximity of the microspore nucleus to the cell wall limiting spindle extension at the GC pole (Brumfield, 1941; Brown and Lemmon, 1991d).

6.5.2 Spindle position

Early evidence for the importance of spindle position for asymmetric division comes from the observation that in triploid *Allium* sp. spores are frequently observed where the spindle is central resulting in symmetric division (Brumfield, 1941). Treatment of microspores with microtubule (MT) poisons or cold temperature also leads to the formation of a central spindle and symmetrical divisions (Terasaka and Niitsu, 1987). The processes which initiate microtubule organizing centre (MTOC) activity and the origin of the spindle poles are poorly understood in plants. Given the absence of structurally distinct centrosomes in plants and the observation that isolated nuclei can act as MTOC *in vitro* (Stoppin et al., 1994), the nuclear membrane may contain a diffuse site with MTOC activity.

It follows then that the position of the microspore nucleus would dictate the site of spindle assembly.

6.5.3 Spindle pole axis

The orientation of the spindle axis has a key role in determining division asymmetry. Centrifugation, cold and caffeine treatments were used to reposition the nucleus in *Pinus* microspores (Terasaka and Niitsu, 1987). Repositioning led to symmetric divisions resulting from a centrally located spindle, and symmetrical divisions in which both daughter nuclei were positioned near the wall. The second type of symmetrical divisions are consistent with an asymmetric spindle position with its axis parallel to the microspore wall rather than perpendicular to the wall which normally occurs. Clearly both spindle position and orientation need to be tightly controlled to generate asymmetric division.

What then triggers the assembly of the spindle and defines the axis of the spindle poles? Brumfield (1941), observed that the appearance of the vegetative spindle pole preceded that of the generative pole, and postulated that this was the result of the reduced amount of cytoplasm at the generative pole, caused by the flattening of the prophase nucleus against the wall. Extending this postulate, the greater concentration of cytoplasmic factors, including tubulin, required for spindle assembly at the inner face of the nucleus would lead first to the inward and central positioning of the VC spindle pole. Fixation of the spindle axis perpendicular to the microspore wall could then be determined by factors which complete spindle bipolarity. Recent studies of the localization of spindle assembly in *Tradescantia* have confirmed that the spindle begins to construct itself in the vegetative pole and that all of the available tubulin in the form of MTs are localised in the cytoplasm at the vegetative face of the nucleus immediately before spindle assembly (Terasaka and Niitsu, 1990; 1995). Further, evidence for cytoplasmic polarity in the premitotic microspore is provided by studies which show that plastids, mitochondria and dictyosomes are excluded from the GC domain and appear to migrate into the vegetative cytoplasmic domain (Brown and Lemmon, 1992).

6.5.4 Nuclear migration

Given the role of the nucleus in dictating the site of spindle assembly, a key process in generating polarity is the control of nuclear migration to the wall. Two factors appear to be important in controlling nuclear migration; vacuolation and the appearance of stage-specific MT arrays. In the majority of species examined a single large vacuole is formed as a result of the fusion of smaller vacuoles, during which time the microspore nucleus migrates to a site adjacent to the wall. In species such as *Tradescantia* (Terasaka and Niitsu, 1990) and tobacco (Twell et al., 1993; Eady et al., 1995) a second nuclear migration occurs to the future GC pole.

This second nuclear migration (NM-2) has been shown to be sensitive to colchicine whereas the first (NM-1) is insensitive (Terasaka and Niitsu, 1990). In these species simple physical growth of the vacuole may account for NM-1 while NM-2 is dependent on a microtubule system which develops after formation of the single large vacuole.

In *Tradescantia* MT arrays appear to dynamically shift over the nucleus during NM-2 as a perinuclear cap at the future vegetative pole. In species such as *Brassica napus* (Hause et al., 1991) and *Arabidopsis* (Owen and Makeroff, 1995) which possess rather spherical microspores it has not been possible to determine whether two nuclear migrations occur. Here the nucleus could arrive at the microspore wall simply as a result of physical displacement by the growing vacuole, or by a more directed MT-dependent transport system. MT localization studies in *Brassica napus* showed that perinuclear arrays and cortical MT bundling were observed to develop in association with nuclear migration (Hause et al., 1991).

Evidence for a specialized MT system that may be involved in nuclear migration comes from studies of orchid species. In *Phalaenopsis* a specialized MT array termed the generative pole microtubule system (GPMS) appears at the future GC pole prior to nuclear migration (Brown and Lemmon,1991c, 1992). Here vacuolation does not occur so nuclear migration may be directly controlled by interactions of the nuclear associated proteins with the GPMS. It is therefore likely that MT-binding proteins such as the MT-dependent motor proteins kinesin or dynein, which have recently been identified in pollen tubes (Tiezzi et al., 1992; Moscatelli et al., 1995, 1996; Cai et al., 1996) are involved in executing microspore nuclear migration.

Further evidence for the role of MTs in the positioning of the premitotic nucleus comes from studies of premitotic cultured tobacco BY2 cells in which the nucleus migrates from the periphery to the central region of the cell. Treatment with cytoskeletal destabilizing drugs showed that premitotic nuclear migration was dependent on MTs (Katsuta et al., 1990). Similarly, Mineyuki and Furuya (1986) showed that it is the MTs that participate in nuclear positioning in the premitotic protonema tip cells of *Adiantum*. Although nuclear migration in premitotic cells is commonly dependent on MTs, maintenance of nuclear and spindle position in BY2 cells also involves actin microfilaments (MFs). Earlier studies of MFs in microspores clearly showed that MFs were associated with the margins of the growing cell plate at PMI but only diffusely staining patterns were observed at earlier stages (Hause et al., 1991). Recently, by applying improved immunolocalization procedures and the anti-MF drug cytochalasin D MFs were shown to be necessary for maintenance of nuclear position in cultured *Brassica* microspores (Gervais et al., 1994). MFs initially radiated out from the centrally located microspore nucleus but became asymmetrically localized to the vegetative face of the acentric nucleus. However, MTs have also been strongly implicated in the maintenance of the acentric microspore nuclear position (Tanaka and Ito, 1981;Terasaka and Niitsu, 1990) and have been observed tethering the nucleus to the plasma membrane (Hause et al., 1991; Brown and Lemmon, 1991c, 1992).

Therefore both components may contribute to the maintenance of asymmetric nuclear position.

6.5.5 Polarity determinants

Although further detailed structural and functional studies of the MT and MF organization need to be carried out it is clear that MT-dependent systems play a key role in establishing the ability of the nucleus to migrate. As such the organization of such MT-systems must respond to a polarity signal or determinant within the cytoplasm. This polarity signal then could be a cytoplasmic gradient of a factor such as a microtubule motor or a regulator of MT assembly localised at the GC pole. Such a gradient could in turn be established by the localization of mRNA encoding such factor at the GC pole. There is considerable evidence for the determination of protein gradients by the MT directed transport or translation of mRNA species in different systems (Evans et al., 1994; Pokrywka and Stephenson, 1995; Kloc and Etkin, 1995). For example mRNA encoding the cell polarity determinant BICOID is localised at the anterior pole of the *Drosophila* embryo through association of the STAUFEN protein with the 3'-UTR of BICOID to form particles which move in a MT-dependent manner (Ferrandon et al., 1994). Evidence for directed mRNA transport in plants has been obtained in developing *Fucus* embryos which show MF dependent-actin mRNA localization at the thallus pole of the polarized zygote followed by its redistribution to the first cell plate by a mechanism insensitive to cytoskeletal inhibitors (Bouget et al., 1996).

6.5.6 The origin of polarity determinants

This leaves the important question of the origin and nature of the signals which execute early events in generating polarity. First, is it possible that polarity may be generated during microspore development by external signals? If external signals operate then such signals could not be polarized with regard to the anther locule since in most species microspores are often randomly orientated with regard to the anther locule and tapetum. However a localised receptor, for example at the future GC pole could act as a polarizing factor if it were localized in the microspore wall, plasma membrane or in a cortical cytoplasmic domain. The activation of such a localised receptor by external conditions in the anther locule could then initiate a signal transduction cascade leading to polarization of the MTs and nuclear migration.

Studies of cell fate determination in *Fucus* zygotes have recently shown that cell fate determinants are indeed localized in the cell wall and act to maintain the differentiated state of thallus and rhizoid cells following asymmetric division (Berger et al., 1994). Cell polarization in animal cells and yeast commonly involve reorganization of both MFs and Mts which are regulated by a hierarchy of GTPases and protein kinases (Glotzer and Hyman, 1995). For example the *PAR1* gene which is required for polarization in the *C. elegans* zygote and early embryo encodes a serine/threonine protein kinase which becomes asymmetrically distributed to a cortical domain before first zygotic division (Guo and Kemphues, 1995). Similar signalling cascades can be envisioned to operate to signal polarity and establish MT reorganization in the microspore.

The localization of such a polarity receptor could be an early event generated during meiosis or tetrad formation or it could be generated during microspore development as a result of intrinsically determined events associated with early gametophytic gene expression. With regard to the first possibility, cytokinesis following meiosis in most dicot species is simultaneous with all intersporal walls developing together. This results in the quadripartitioning of the postmeiotic cytoplasm into four equal "spore domains" (Brown and Lemmon, 1991a,b and references therein). Intrinsic polarity within the spore domains, which could ultimately direct nuclear migration, could be generated as part of the process of generating the spore domains.

Detailed studies of the cytoskeleton and phragmoplast development have revealed radial arrays of MTs emanating from the postmeiotic nuclei which interact and define the boundaries of the spore domains and the sites of phragmoplast development (Brown and Lemmon, 1991a,b). The localization of actin microfilaments in the multipolar spindle and MFs radiating from the four nuclei also indicates their involvement in phragmoplast development (Van Lammeren et al., 1989; Schopfer and Hepler, 1991). Together, such systems could establish polarity by generating a cortical cytoplasmic domain or by delivering polarity determinants to membrane or wall locations as discussed above. Direct evidence for early cytoplasmic polarity is provided by observations which show that future apertural sites are defined by the presence of endoplasmic reticulum (ER) closely associated with the plasma membrane in the tetrad. These polar ER configurations reflect the tetrad cleavage planes and appear to be dependent upon spindle orientation at meiosis I and II, since disruption of meiotic spindles with colchicine and/or centrifugation prevents aperture formation or leads to apertures in incorrect positions often with unusual patterning features (reviewed in Heslop-Harrison, 1971).

In species which shed pollen in permanent tetrads it has been possible to determine the polarity of nuclear migration. Nuclear migration can either be towards the inner (radial) or to the outer (distal) wall of the tetrad, but is usually constant within a species (Geitler, 1935). So although the axis of polarity may differ between species, which may suggest diversity in the processes which generate polarity, nuclear migration appears to be predetermined towards a fixed site. This raises the question of whether the ability of the microspore to achieve asymmetric division is dependent upon the nucleus arriving at this site? Asymmetric division in *Pinus* microspores was shown to occur at any site along the radial or distal walls following nuclear displacement by centrifugation or cold treatments (Terasaka and Niitsu, 1987). Therefore, the formation of an asymmetric spindle and asymmetric division are not dependent on arrival of the nucleus at a particular location. The proposed role of the system generating regular polarity then is first, to ensure nuclear migration to the wall and second, to ensure that the correct spindle axis is established to achieve asymmetric division.

6.6 Gametophytic mutations affecting pollen development

The first mutations in gametophytically expressed genes were reported over 70 years ago by Parnell (1921) who identified phenotypic segregation following iodine staining in pollen of rice plants heterozygous for the *glutinous* character of the endosperm. Similar observations were made by Brink and MacGillivray (1924) for the corresponding waxy character of maize pollen and endosperm, who further observed reduced pollen transmission of both *waxy* and *glutinous* alleles, and hypothesized, in referring to the growth of the pollen tube, that 'this presumably is controlled by the action of certain factors active in the tube nucleus'. These genetic studies provided the first evidence for the important postmeiotic function of genes in the male gametophyte. The identification of null alleles corresponding to biochemical deficiencies such as for alcohol dehydrogenase activity in maize pollen (Freeling, 1976; Stinson and Mascarenhas, 1985) and ß-galactosidase activity in *Brassica campestris* pollen (Singh et al., 1985) has provided further genetic evidence of gametophytic gene expression.

The first gametophytically acting pollen lethal gene was described in Zea mays (Singleton and Mangelsdorf, 1940). This gene, termed *small pollen*-1 (sp), resulted in a phenotype in which two pollen sizes, large and small, were produced in approximately equal numbers in plants heterozygous for sp. The common observation of the lack of pollen transmission, and the associated segregation for a similar *small pollen* phenotypes in nullisomic and chromosome region-specific deficient lines of maize, indicates that important genes required for male gametophyte development are both numerous and dispersed throughout the genome (Stadler, 1933; Stadler & Roman, 1948; Zhao and Weber, 1989). In a recent survey of pollen abnormalities in hypoploid plants derived from 26 B-A translocations in maize, a minimum of 23 distinct factors leading to developmental delay and/or arrest at the uninucleate or binucleate stages were identified (Kindinger et al., 1991).

Distorted Mendelian segregation ratios which specifically affect the transmission of particular alleles through pollen have been detected repeatedly, and include examples where such genes affect pollen development, pollen germination or tube growth (reviewed in Ottaviano and Mulcahy, 1989). Early examples were noted in *Melandrium, Datura* and *Hordeum* (Jones, 1928). In *Medicago falcata* and lima beans preferential fertilization was shown to be the consequence of unequal pollen tube growth (Bemis, 1959). A pollen killing factor (ki) which caused microspore abortion was detected in wheat (Loegering and Sears, 1963), while a distinct pollen killer factor was detected in *Nicotiana tabacum*, which was carried by an alien chromosome of *N. plumbaginifolia* (Cameron and Moav, 1957). This factor, in contrast, caused abortion in pollen not carrying the alien chromosome. Deleterious alleles of genes which presumably affect haploid gene expression also include chlorophyll mutants in barley which showed distorted segregation ratios and reduced viability of microspores carrying mutant alleles (Doll, 1967). Other studies have shown reduced pollen transmission due to slower germination of pollen carrying the recessive alleles waxy (Sprague, 1933) and *opaque*-2 in maize (Sari Gorla and Rovida, 1980). In maize nine different gametophytic factors have been described which confer high competitive ability to pollen tubes carrying mutant ga alleles (Bianchi and Lorenzoni, 1975). In contrast a class of maize gametophytic mutants, *defective endosperm (de)*, negatively affect pollen tube growth and block endosperm development (Ottaviano et al., 1987). A similar gametophytic/sporophytic overlap was noted for two embryo lethal mutants in *Arabidopsis* which showed reduced male gametophytic transmission (Meinke, 1982; Meinke and Baus, 1986). Using an alternative antisense mutational approach the tomato *lat52* gene was shown to be an essential gametophytic gene required for correct pollen germination and tube growth (Muschietti et al., 1994). Recently, a male gametophyte-specific mutation has been identified in maize, GAMS-1, which led to the arrest of pollen development at the bicellular stage (Sari-Gorla et al., 1996).The above examples illustrate that it is possible to identify diverse gametophytically-acting mutations. Therefore systematic mutant screening may be a particularly valuable approach to identify gametophytically expressed genes required for the development and function of the male gametophyte.

Evidence for gametophyte-specific mutations in *Arabidopsis* is very limited. Redei (1965) described a mutation, *gametophyte factor 1* , which mapped to upper arm of chromosome 2, whose transmission was completely blocked through the female but showed reduced (32%) transmission through pollen. More recently Springer et al. (1995) identified a transposon-mediated "gene-trap" mutant termed *prolifera* which showed reduced (50%) transmission through the female with only slightly reduced transmission through pollen. The *PROLIFERA* gene was shown to be related to the MCM2-3-5 family of proteins which are required for the initiation of DNA replication in yeast. This suggests that DNA replication during male and female gametophyte development may be controlled by different genes. Using an alternative mutational approach the *Brassica napus Bcp-1* gene was shown to be essential for correct maturation of pollen in *Arabidopsis* (Xu et al., 1995). Antisense downregulation of *Bcp-1* driven by the pollen-specific *lat52* promoter showed that *Bcp-1* acts gametophytically after PMI. Furthermore, antisense downregulation of *Bcp-1* in the tapetum demonstrated that *Bcp-1* also has an essential sporophytic role earlier during microspore development.

The first gametophytic mutant of *Arabidopsis* shown to specifically affect male gametophyte development has been termed *sidecar pollen* (*scp*; Chen and McCormick, 1996). Plants heterozygous for the *scp* mutation show 50% aborted pollen in the Landsberg Erecta background but in the Nossen background 48% of pollen are aborted and 7% are shed in a divided condition with an extra uninucleate vegetative-like cell. This extra "vegetative" cell is attached to a cell containing a vegetative-like nucleus and two sperm-like nuclei. Other than being joined the extra vegetative cell appears to mature relatively normally in that the *lat52-gus/nia* (Twell et al., 1994) and lat59-gus (Twell et al., 1990) markers are activated and germination occurs *in vitro*. This mutant appears to undergo an extra symmetrical cell division at the microspore stage and it is proposed that this may result either from the mis-orientation of the spindle apparatus during PMI or the premature division of an unpolarized microspore. Therefore *SCP* may encode a component required for nuclear migration and/or asymmetric spindle orientation, or a repressor of cell division normally active until the microspore achieves polarity. Further detailed analysis of the precise time of production of the extra cell in *scp* mutants should shed light on these hypotheses and why *scp* carrying pollen frequently aborts. SCP has been mapped to chromosome 3 near the markers nga6 and BGL1.

6.7 Sporophytic mutants as tools to analyze microspore polarity and the action of gametophytic mutations

Two recently isolated sporophytic mutants in *Arabidopsis* affecting male gametophyte structure and development are described below which have particular value as tools for the analysis of microspore polarity and male gametophytic mutants.

6.7.1 *quartet*

Two mutant loci have previously been reported, *quartet 1(qrt1)* and *qrt2* which shed pollen in a tetrad configuration, wherein all four meiotic products remain joined in a characteristic tetrahedral array (Preuss et al., 1994). The outer walls of the four meiotic products of the meiocyte are fused, which in qrt1 is linked to a reduction in callose production separating the young microspores at the tetrad stage. Despite this early defect the cytoplasmic contents of each microspore remain discrete and pollen matures correctly. For example, in single tetrad pollinations each of the four pollen grains are capable of fertilization to produce four viable seed. Furthermore in plants hemizygous for introduced *lat52-gus* gene fusion in the *qrt1/qrt1* background, only two of the four members of the tetrad stain positively for GUS activity. The *quartet* phenotype is particularly valuable in characterizing male gametophytic mutants since in a qrt background gametophytic mutants will show the phenotype in two of the four pollen grains within a tetrad

(Chen and McCormick, 1996). The *qrt* mutants may be of further value in determining whether microspore nuclear migration occurs towards a predetermined site in *Arabidopsis* since the four meiotic products remain attached apparently in a fixed tetrahedral arrangement. Further applications of the qrt mutants in genetic mapping and the analysis of mechanisms involved in the evolution of species with multiple joined tetrads or "polyads" have been discussed (Smyth, 1994). In the screen described below (Section 6.8) we have isolated 7 independent mutants which show the *quartet* phenotype out of a population of 10,000 M2 individuals. At this stage it is unknown whether these newly identified fused pollen grain mutants are allelic to *qrt1* or *qrt2* or represent new *qrt* loci. *qrt1* maps to chromosome 5 (position 84 ± 1.5) and *qrt2* to chromosome 3 (position 3 ± 2).

6.7.2 *stud/tetraspore*

A number of reduced fertility mutants of *Arabidopsis* which result in the formation of large multicellular pollen grains have recently been reported. These mutants have been independently classified as two allelic series in genes termed *STUD* (Hulskamp et al., 1997) and *TETRASPORE (TES;* Spielman et al., 1997). *stud/tes* mutations are all sporophytic monogenic recessive and prevent the formation of separating callose walls after meiosis II. This initially creates coenocytic microspores (CM) composed of four nuclei within a common cytoplasm. Similar mutants have been described in soybean (*male sterile-1(ms1)*; Albertsen and Palmer, 1979) and alfalfa (*jumbo pollen* (jp); McCoy and Smith, 1983). However the subsequent development of CMs differs between these and the *Arabidopsis* large multicellular pollen mutants. In *ms1* microspore nuclei do not fuse and CMs usually degenerate after accumulation of lipid and starch. In *jp* postmeiotic fusion of all four nuclei occurs frequently (80-95%) resulting in mature 4N pollen grains containing a single vegetative and generative nuclei. Incomplete nuclear fusion before PMI appears to occur in the remaining CMs (5-20%) resulting in mature pollen grains with additional vegetative or generative like nuclei. In the *Arabidopsis* mutants however nuclear fusion is far less frequent and up to four sperm cell pairs have been observed in mature pollen grains suggesting that the four nuclei can each undergo nuclear migration and asymmetric division within the common cytoplasm. The *jp* and *Arabidopsis* mutants demonstrate that cytokinesis and microspore isolation after meiosis II are not prerequisites for microspores to achieve polarity (nuclear migration and asymmetric division), and suggest that polarity determinants can operate locally and independently within each spore domain. Given the absence of radial cell walls at meiosis, if polarity determinants are established early during the generation of the spore domains as discussed above (Section 5.6), these would have to be associated with the distal wall or a cortical cytoplasmic domain.

Four large multi-nucleate pollen mutants with similar phenotypes to *stud/tes* mutants were identified in the screen described below (Section 6.8) and were shown to be allelic to *tes* (Howden and Twell, unpublished). Both STUD and TES map to the lower arm of chromosome 3 close to gl-1 and therefore are likely to represent the same gene.

The STUD/TES product is clearly required for the construction of the dividing walls following meiosis II. The gene product could conceivably be involved in some aspect of targeting of wall precursors to the sites of wall ingrowth and therefore may be a component of the vesicle transport docking or fusion apparatus. Other sporophytic mutants which act in the early embryo and developing seedling have been identified which affect cell plate synthesis. These include the "pattern formation" mutants *gnom/emb30* (Mayer et al., 1993), *fass* (Torres Ruiz and Jurgens, 1994) and *knolle* (Lukowitz et al., 1996). Both the *gnom* and *knolle* mutants result from defects in putative components of the secretory apparatus. The GNOM protein shows sequence similarity to yeast Sec7 which is involved in protein transport through the golgi (Shevell et al., 1994), while KNOLLE is similar to T-SNARE proteins which are involved in vesicle docking (Lukowitz et al., 1996). Both mutants disrupt the normally regular planes and symmetry of division in the developing embryo and the *knolle* mutant in particular shows cells with partial walls and multi-nucleate cells which appear to result from incomplete cytokinesis. The similarity in the site of action of these mutants suggests that the STUD/TES gene could encode a related protein which is specifically expressed during male meiosis.

Mutants at the *STUD/TES* locus may find application in the analysis of the action of gametophytic mutations. If male gametophytic lethal *(mgl)* mutants are completely penetrant it will not be possible to isolate homozygous individuals and therefore test whether gametophytic mutants also have a sporophytic phenotype. In the *stud/tes* mutant background however the lethal mutation may be complemented in the common cytoplasm by the product of the wild type allele (assuming a diffusible product) and it would be possible to isolate a homozygous *(mgl/mgl)* individual. If however the presence of the wild type product did not complement the mutant phenotype this would suggest that these gametophytic factors act locally in spore domains within the common cytoplasm.

6.8 Screening for male gametophytic mutants

Given the success of the genetic approach in identifying molecular components of the polarity generating mechanisms and cell fate determinants in other systems (Horvitz and Herskowitz, 1992; Rhyu and Knoblich, 1995; Kraut et al., 1996) this should be a powerful approach for the dissection of microspore polarity and determination. Mutants which show more symmetrical divisions at PMI are expected to result from lesions in components of the polarity signalling mechanism or in multiple components required for the expression of polarity as detailed above (Section 6.5). We are using several molecular-genetic screens to isolate gametophytic mutants using *Arabidopsis thaliana*. These screens have led to the isolation of several classes of novel gametophytic mutants which independently affect microspore viability, microspore division symmetry and generative cell development.

6.8.1 Direct visible screening

Approximately 10,000 M2 plants from an EMS mutagenised population (Nossen (No) background) were screened for visible male gametophytic mutants by examining mature pollen grains after DAPI staining using fluorescence microscopy (Fig. 6.3). At maturity *Arabidopsis* pollen is tricellular containing with a large vegetative nucleus and two smaller sperms cells with highly condensed intensely staining nuclei (Fig. 6.4). M2 seedlings were planted in a 96 "microtitre plate" arrangement and at anthesis mature pollen was collected from each plant in a microtitre plate and stained with DAPI. DAPI stained pollen was initially examined at low magnification to detect obvious morphological including *gemini* (divided or twinned) pollen mutants. Examination at higher magnification under UV illumination was then used to identify any mutants affecting cell division from the number or characteristic nuclear organization of the male germ unit.

Approximately 500 EMS induced mutants were isolated and these form the basis of a database of mutants affecting pollen development. A wide range of pollen mutants were detected in the screen, several of which include phenotypes not previously described. Putative mutants were examined for other visible morphological mutations in the sporophyte and M3 seed regrown to determine whether the mutation was transmitted to the next generation. At this stage it was observed that a number of putative mutants appeared to segregate for other recessive phenotypes in the M3 population; these frequently included pale or albino seedlings. In order to "cleanse" the genetic background of additional sporophytic mutations induced by the EMS, mutants were backcrossed to the parental No ecotype and the male and female transmission of the mutant phenotypes determined.

Male gametophytic lethal (*mgl*) mutants. As expected in such a screen, a high frequency (~1%) of putative male gametophytic lethal mutations were detected in which approximately 50% of the pollen grains were aborted (Fig. 6.4). This phenotypic class is likely to mutations in a diverse array of genes required for basic microspore cell function and metabolism. The size and apparent homogeneity of this major class makes it difficult to gain insight into the role of such genes without further detailed phenotypic analysis. Perhaps the best way to make progress in understanding the role of such genes in microspore development is to use gene-tagging strategies (Section 6.8.2) to directly clone such genes though the identification of "tagged" *mgl* mutants. In this way clues as to the identity of the protein(s) involved may be obtained from sequence comparisons and functional analyses of the localization and properties of protein(s) in wild type and mutant pollen.

Gametophytic cell division mutants. As predicted in such a screen a range of mutant phenotypes were observed that affected the stereotypical pollen cell divisions. These division mutants which were all represented by multiple independent isolations were classified into three groups based upon the proportion of pollen grains showing particular phenotypes. All three groups contain mutants resulting from gametophytically acting mutations since: (1) plants are fertile with normal seed set; (2) phenotypes are observed in 50% or less of the pollen

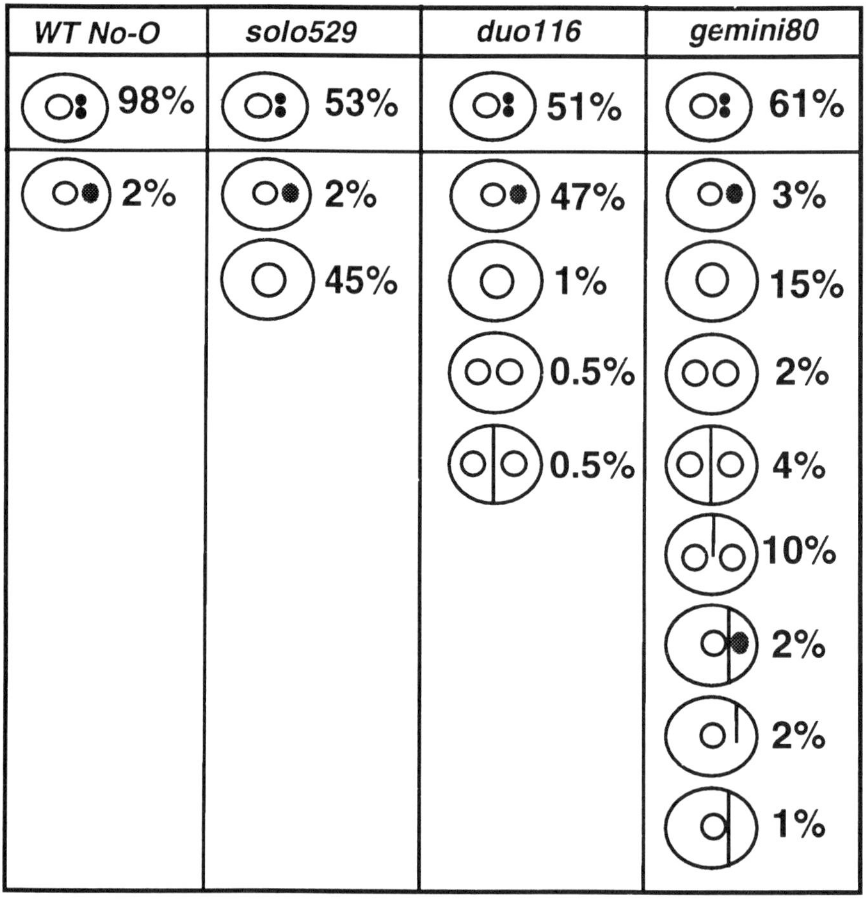

Fig. 6.3. Flow diagram showing steps involved in direct visible screening for male gametophytic mutants and their further characterization aimed towards map-based cloning.

population; (3) heterozygous individuals show the phenotype; (4) transmission of the mutant allele is blocked or reduced through the pollen parent; and (5) the phenotype is observed in a maximum of two of the four members of a tetrad when the heterozygous mutant is present in a *qrt/qrt* background.

Solo mutants (unicellular pollen, failing to divide at PMI). In *solo* pollen mutants approximately 50% of mature pollen grains contain a single nucleus, indicating

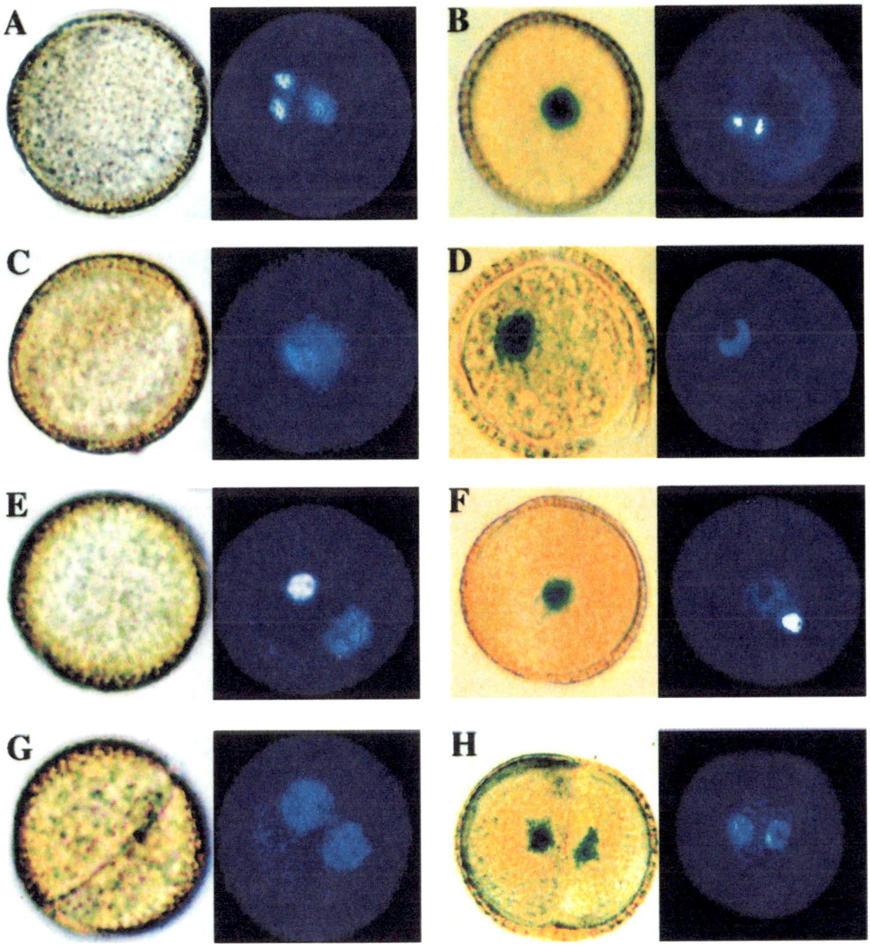

Fig. 6.4. Photographic images of mature pollen isolated from wild type and gametophytic mutants of Arabidopsis. Bright field and corresponding DAPI stained epifluorescence images are shown in the left and right panels of A-H respectively. A, B wild type pollen; C, D solo mutant; E, F duo mutant; G, H gemini mutant. B, D, F and G show nuclear-localized GUS staining in wild type, solo, duo and gemini mutants respectively containing the lat52-gus/nia VC fate marker.

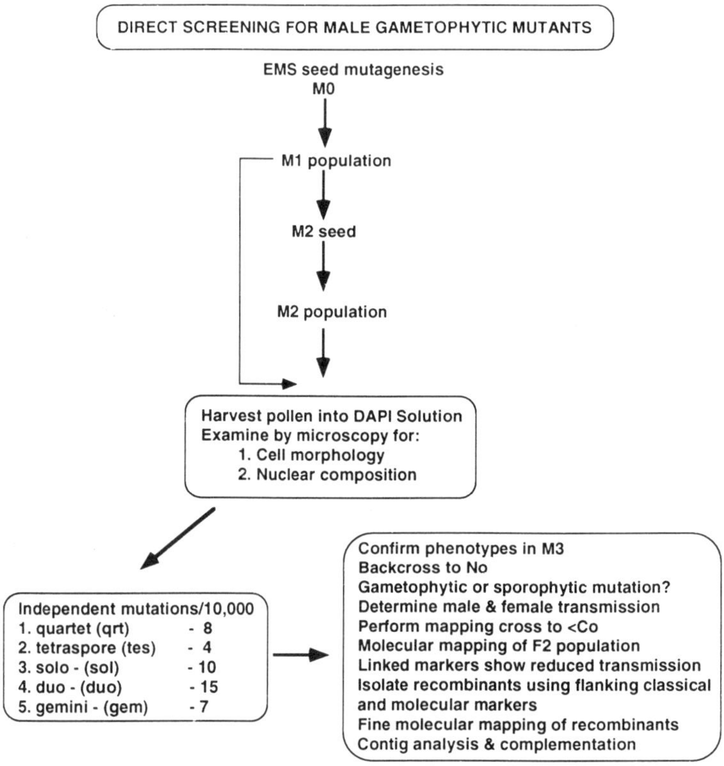

Fig. 6.5. Diagram showing the frequencies of aberrant pollen phenotypes observed in wild type and a representative of each of the phenotypic mutants classes solo, duo and gemini.

failure of cytokinesis at PMI (Figs. 6.4 and 6.5). These mutations which affect the ability of the microspore to divide may therefore affect a structural or regulatory component in the cellular division process. It is unlikely that these mutants affect components only involved in cell plate synthesis and formation since if microspores completed nuclear division then binucleate pollen grains would result. These have only infrequently been observed in the gemini mutants described below. It is more likely that *solo* pollen results from a failure to enter mitosis. As such these mutants may be thought of as cell cycle mutants in that they affect the ability of the microspore to complete its cell cycle. Components of the spindle apparatus or including factors regulating the assembly and/or function of the microtubule system are putative candidates. It is interesting to note that *solo* pollen grains tend to be slightly larger than wild type grains at maturity indicating that the cells may have undergone DNA synthesis but have failed to enter mitosis. Further studies of the nuclear migration and progression of the cell cycle using microdensitometry (Binarova et al., 1993; Carmichael and Friedman, 1995) in wild type and *solo* microspores should lead to further insight into the effect of these mutations. Whether *solo* mutants express polarity and show normal nuclear migration remains to be determined. However, if *solo* mutants also fail to express polarity *SOLO* genes may provide a link between the expression of cell polarity and cell cycle progression.

Despite the possession of a single nucleus *solo* microspores undergo maturation including cell expansion and the synthesis of a dense cytoplasm (Fig. 6.4). The GUS positive staining of all pollen grains in *solo* mutants carrying the *lat52-gus* gene in the background shows that *solo* grains correctly activate the VC-specific promoter *lat52* (Fig. 6.4).Therefore in solo pollen the single nucleus adopts the fate of the vegetative nucleus of wild type pollen grains. Similarly, if division at PMI is directly blocked with colchicine tobacco microspores express *lat52-gus* when cultured under maturation conditions (Eady et al., 1995). These data confirm that VC fate is the default condition of the microspore nucleus in the absence of division at PMI.

Duo mutants (bicellular pollen, failing to divide at PMII). The *duo* pollen phenotype (Fig. 6.4) is indicative of pollen which has failed to enter or complete division at PMII. In some cells the chromosomes appear to have initiated condensation in prophase but arrested prior to alignment on the metaphase plate. In the *duo116* mutant approximately 50% of the pollen population shows the *duo* phenotype consistent with a gametophytically acting mutation (Fig. 6.5). These phenotypes suggest that the mutation may be a structural or regulatory component required for cell division, but that such components would specifically function in the symmetrical second division of the generative cell. *DUO* genes may therefore be specifically expressed in the GC, or encode proteins whose function is essential for completion of the GC cycle. Such proteins would not function in the microspore or female gametophyte and so could represent specialized isoforms of cell division factors. The low frequency of symmetrically divided pollen grains observed in the *duo116* mutant (Fig. 6.5) suggests that *DUO116* may also participate in regulating division asymmetry at PMI. Incorporation of the *lat52-gus/nia* marker into the *duo116* mutant showed that the VC correctly activates *lat52* in the absence of a functional GC and that GC fate is unaltered in that the arrested GC correctly fails to activate *lat52* (Fig. 6.4)

Gemini mutants (divided, twin-celled pollen, more symmetrical division at PMI). The *gemini* class of mutants display the widest range of pollen phenotypes which are illustrated in Fig. 6.5. In *gemini* mutants the symmetry of division at PMI is affected such that symmetrical and partial divisions frequently occur (Figs. 6.4 and 6.5). In symmetrically divided cells both daughter cells correctly activate the *lat52-gus/nia* marker and show dispersed nuclear chromatin (Fig. 6.4) in accord with previous observations which demonstrate that induced symmetrical divisions result in the default VC fate in both daughter cells (Eady et al., 1995). Initial characterization of the *gemini* mutants by aniline blue staining reveals that the persistent cell walls present at pollen maturity are composed of callose and often are seen to connect with the pollen grains wall even if these walls are incomplete. Such phenotypes may be expected if the spindle position or axis is altered.

Normally the microspore nucleus at incipient PMI is positioned against the microspore wall. The spindle axis is set up perpendicular to the wall such that the cell plate is initiated parallel to the outer wall and is subsequently connected to the outer wall by centrifugal growth of the cell plate at its margins (Hause et al., 1991; Owen and Makeroff, 1995). This appears to occur though the delivery of vesicles containing wall precursors and the callose synthetic machinery. However, if the spindle axis was rotated 90° in *gemini* mutants and therefore parallel to the microspore wall, delivery of vesicles and the centrifugal growth of the phragmoplast would connect the cell plate to the outer wall. The inward growth of the cell plate would most likely be inefficient and fail. MTs have been observed connecting the microspore nucleus to the plasma membrane when the nucleus is asymmetrically positioned near the wall (Hause et al. 1991; Brown and Lemmon, 1991c, 1992). As discussed above (Section 6.5.2) if MTOCs are associated with the nucleus and serve to anchor the spindle poles, spindle pole position may be influenced by interactions between the nucleus and cortical domains or factors linked to the plasma membrane.

A further interesting phenotype in the *gemini* mutants is that nuclear division (karyokinesis) is often uncoupled from cellular division (cytokinesis). This commonly results in a smaller enucleate portion of the microspore cytoplasm being walled off from a larger cell containing one or two nuclei (Fig. 6.5). These "uncoupling" phenotypes demonstrate that although cytokinesis and nuclear division are normally closely coupled these processes can operate independently. This concept has been elegantly demonstrated in experiments in which chromosomes were physically removed from the mitotic spindle of grasshopper spermatocytes. Chromosomeless cells still executed normal cytokinesis to produce enucleate daughters demonstrating that cleavage can operate independently of chromosomes (Zhang and Nicklas, 1996). *GEMINI* genes may therefore function as checkpoint genes ensuring close coupling of karyokinesis and cytokinesis. *gemini* mutants may therefore be of particular value in understanding how karyokinesis and cytokinesis are normally coordinated during asymmetric cell division.

It is also possible that *gemini* phenotypes result directly from defects in components of the intracellular targeting machinery required for the asymmetric positioning of the cell plate. As discussed above (Section 6.7.2) the gnom and knolle mutants alter the position and correct synthesis of the cell plate and which results from defects in putative components of the secretory apparatus. If similar proteins function in the positioning and synthesis of the cell plate at PMI it is possible that *GEMINI* genes encode gametophyte-specific versions of these proteins.

It is apparent that *GEMINI* genes could encode proteins which function at several points in the pathway which signals and executes microspore polarity. Further detailed phenotypic analysis and the molecular cloning of *GEMINI* genes is now required to begin to define the role and identity of molecular components comprising this pathway.

Allelism and map-based cloning. The isolation of multiple gametophytic mutants with similar phenotypes, and multiple alleles of sporophytic genes affecting pollen development suggests that saturation for gametophytic mutants affecting division polarity at PMI was not far away in our M2 population. However, it will be necessary to genetically determine allelic relationships among all mutants with similar phenotypes to be confident of this conclusion. This raises an important difference in strategy and effort compared with the analysis of allelic relationships among sporophytically determined mutations. For most recessive sporophytic mutations simple complementation tests can be performed by crossing two independent mutants together and observing the phenotypes of the F1 progeny. Generating such F1 progeny carrying two independent male gametophytic mutations is often not possible since these mutations commonly block male gametophyte function. For incompletely penetrant alleles at a given locus it would be possible to confirm allelism if the proportion of aberrant pollen grains in F1 progeny resulting from each allele were additive. Mapping each gametophytic mutation would provide definitive data if mutations mapped to unlinked loci. However, non-allelic mutations may still map very closely together. Definitive data would only be made possible following the cloning of one of the genes, enabling sequence comparison of the same gene in independently isolated mutants.

To determine allelism and initiate cloning of the genes involved, the *gemini* mutations are being mapped using molecular markers. Multiple microsatellite primers currently exist which show useful polymorphism between No and other ecotypes (Bell and Ecker, 1994; Konieczny and Ausubel, 1993). Fine mapping studies will require the isolation of further SSLP and CAPS markers showing detectable polymorphisms in particular chromosomal regions.

Normal F2 genetic mapping procedures are slightly modified when dealing with mutations affecting gametophytic development because of reduced gametophytic transmission. This is further complicated with the penetrance factor of the mutation itself. If a gametophytic mutation shows significantly reduced transmission it is possible to detect genetic linkage through the variance of the expected marker segregation in a F2 population. If variance from the expected 3:1 ratio is observed then this indicates linkage with the mutation. Therefore it is possible to detect linkage in an F2 population without actually scoring for the visible gametophytic phenotype, where reduced transmission is scored as the phenotype (McCormick et al., 1994). However, if the F2 plants are also scored for the mutant phenotype it is possible to isolate recombinants between the linked locus and the marker which assists in fine mapping. Initial mapping studies of the *gem80* and *duo116* mutations indicate that they map to different chromosomal locations and therefore affect different genes. With the ever increasing production of overlapping BAC, YAC and cosmid contigs, chromosome walking procedures are now far less time consuming and more predictable (reviewed in Leyser and Chang, 1996).

Targeted transposon-tagging. Once a reasonably accurate map position has been established for a given mutation cloning of the gene responsible may be accelerated through targeted transposon-tagging (for review see Roberts, 1996). Non-targeted or random transposon-tagging of specific gametophytic genes is unlikely to be a realistic approach since the frequency of visible sporophytic mutations is of the order of 100-fold lower than in our EMS population. Assuming the frequency of visible gametophytic mutations are similarly reduced this would mean screening 1,000,000 individuals to obtain the number of mutants observed in our visible screen (Fig. 6.3). This figure may be reduced to approximately 140,000, if for example, all 7 *gemini* mutants were allelic. On the other hand targeted transposon-tagging from closely linked sites, which can increase the tagging frequency by a factor of 100, would mean that only a few thousand individuals may have to be screened to isolate a tagged mutant at a specific locus. These figures may be reduced even further if one considers that individuals within a given population will give rise to multiple mutant sectors with tranposons at different new sites. Screening of flowers derived from, for example, 3 inflorescences would reduce the number of individuals that have to be grown. Targeted-transposon tagging of gametophytic mutants is further simplified since phenotypes resulting from new insertions can be observed directly in the M1, such that it is unnecessary to introduce the transposon into the mutant background. Several groups have mapped a number of active or defective transposable elements on all five chromosomes using the Ac/Ds and En/I element systems (Sundaresan et al., 1995; Aarts et al., 1995). These elements and the ongoing mapping of further elements are likely to provide valuable tools for the accelerated cloning of mapped gametophytic genes.

6.8.2 Screening for T-DNA and transposon-tagged gametophytic mutants by segregation distortion

A potentially rapid and systematic, although labour intensive, approach for the isolation of important genes involved in gametophytic development is provided by random gene-tagging by T-DNA or transposon-mediated insertional inactivation. Since the inactivation of many important genes controlling gametophytic development is likely to lead to reduced transmission through the male or female gametes, it is possible to screen for the reduced transmission of a resistance marker carried by a T-DNA or a transposon. For example, if a single locus insertion carrying a seedling expressed resistance marker inactivates a postmeiotically-expressed gene essential for male gametophytic development, then the ratio of resistant:sensitive seedlings in selfed progeny would be 1:1 rather than the expected 3:1 (Fig. 6.6). This approach initially does not allow one to distinguish between mutations affecting transmission through male or female gametes, but is attractive in that: (1) it will identify important gametophytically expressed genes required for the complete developmental pathway of male and female gametophyte development, including gamete recognition and fusion; (2) genes identified in this way are very likely to be tagged or very closely linked to a molecular maker; and

(3) it is simple to carry out given the availability of large populations of transgenic lines containing a high proportion of single locus T-DNA or transposon insertions.

We have screened approximately 2,000 T-DNA lines and 1,500 transposon containing lines. In both approaches approximately one percent of the lines show segregation distortion (~1:1) for antibiotic resistance. This suggests that both T-DNA and transposon tagging of important gametophytic genes through segregation distortion is similarly efficient depending on the availability or ease with which seed from large numbers of individuals hemizygous for single locus insertions can be generated. Here transposons may have the advantage in that we have observed transposition of single locus elements into new locations at a frequency of ~70%, (Howden, Grossniklaus and Twell unpublished), whereas T-DNA populations rarely show more than 50% single locus insertions (Feldman, 1991). A further consideration is that screening for segregation distortion may not be specific for the

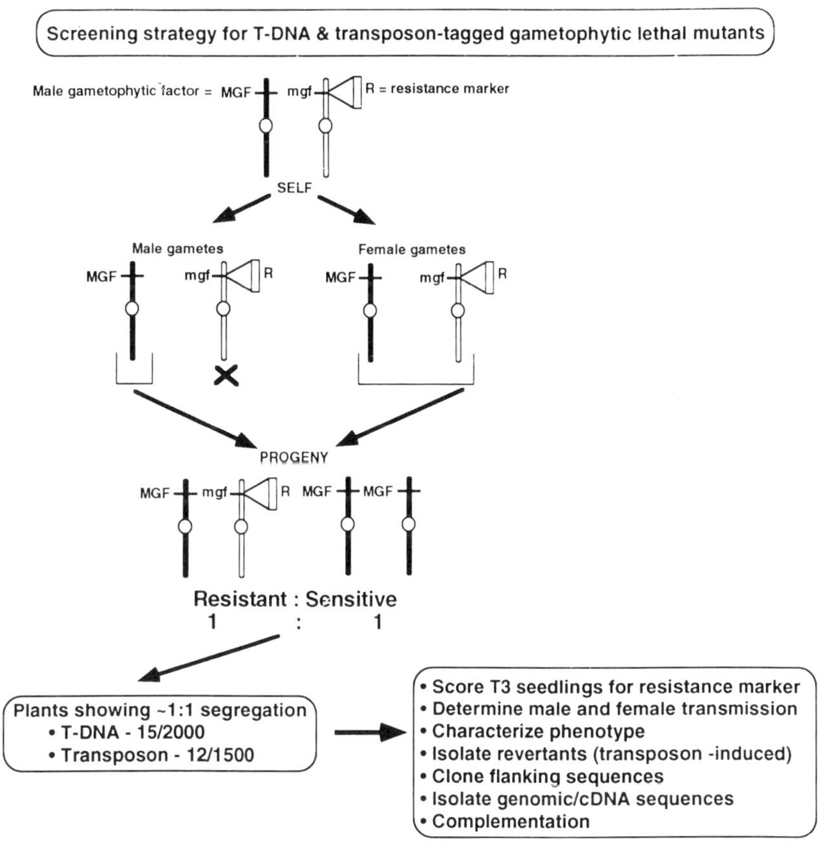

Fig. 6.6. Flow diagram showing a screening strategy for the isolation of T-DNA and transposon-tagged male gametophytic lethal mutations and steps in their subsequent analyses.

isolation of single gene gametophytic mutations. Segregation distortion of resistance markers may also indicate chromosomal rearrangements including inversions, aneuploids and reciprocal translocations which would involve multiple genes (Vizir et al., 1994).

In a proportion (~ 20%) of the T-DNA lines examined, which have shown reliable 1:1 segregation ratios, further analysis has shown that occasionally some siblings are generated which segregate normally (3:1) for the resistance marker. This suggests that the T-DNA may not be the direct cause of the reduced transmission by insertional inactivation, and that T-DNA may have inserted at a site closely linked to a gametophytic mutation. Whether aberrant T-DNA integration events or the transformation procedure are the cause of the gametophytic mutation is not clear. In either case T-DNA provides a start point for chromosome walking towards the gametophytic mutation. Here the resistance marker allows the determination of linkage between the T-DNA and the gametophytic mutation even when the gametophytic phenotype is not easily scored in large populations (such as a reduction in pollen germination or pollen tube growth). For both tagged and T-DNA linked mutations direct access to the gene is most efficiently achieved by the cloning of plant DNA flanking the insertion using Inverse-PCR (Thomas, 1996) or TAIL-PCR (Liu et al., 1995) procedures. If a concerted effort could be made to identify large numbers of random T-DNA or transposon induced gametophytic mutations many genes with important cellular functions could be isolated in a similar strategy to that being developed for assigning the function of embryo lethal mutations through random T-DNA tagging (Errampalli et al., 1991).

6.9 Perspective

The identification and characterization of genes which play essential roles in the control of asymmetric cell division and cell fate in developing pollen would be a major step forward in understanding gametophyte development and the control of cell fate in plants. Several molecular-genetic strategies designed to identify such genes have been evaluated. This has led to the identification of three novel classes of gametophytic mutants which affect cell cycle progression and division asymmetry at PMI. The identification of such mutants means that it is feasible to begin to dissect the complex genetic components which are involved in the control of cell polarity and asymmetric cell fate. Furthermore, the gametophytic screening approaches described are likely to be of wider value since they also provide access to genes involved in the control of many key cellular processes such as cell cycle control, cell wall metabolism, cytoskeleton dynamics and intracellular signalling and transport. With increasing resources directed towards map-based cloning and transposon tagging it is realistic to expect that the molecular cloning of key genes involved in generating microspore polarity and differential cell fate is not far away.

Acknowledgements

We gratefully acknowledge The Royal Society for support to DT and BBSRC under the Cell Commitment and Determination Initiative to RH. Work on screening transposon populations in collaboration with Ueli Grossniklaus was supported by a HFSP fellowship to RH. We also thank George Pelletier, Christine Horlow and Sandrine Bonhomme for their willing collaboration on T-DNA screening, MOGEN and Keith Lindsey for providing access to T-DNA populations, and Hugh Dickinson, Bob Pruitt, Melissa Spielman and Rod Scott for providing preprints and helpful discussions on the *stud/tes* mutants.

References

Aarts MGM, Corzaan P, Stiekema W, Pereir, A (1995) A two element Enhancer-Inhibitor transposon system in Arabidopsis thaliana. Mol Gen Gene 247:555 – 564

Albertsen MC, Palmer RG (1979) A comparative light – and electron-microscopic study of microsporogenesis in male sterile (ms1) and male fertile soybeans (Glycine max (L.) Merr.). Ame. J Bot 66:253 – 265

Bemis WP (1959) Selective fertilization in lima beans. Genetics 44:555 – 562

Bianchi A, Lorenzonz C (1975) Gametophytic factors in Zea mays. In DL Mulcahy(ed) Gamete competition in plants and animals. Elsevier, Amsterdam

Bell CJ, Ecker JR (1994) Assignment of 30 microsatellite loci to the linkage map of Arabidopsis. Genomics 19:137 – 144

Benito-Moreno RM, Macke F, Alwen A, Heberle-Bors E (1988) In-situ seed production after pollination with in-vitro-matured, isolated pollen. Planta 176:145 – 148

Berger, F., Taylor, A. and Brownlee, C. 1994. Cell fate determination by the cell wall in early Fucus development. Science 263, 1421 – 1423

Binarova P, Straatman K, Hause B, Hause G, van Lammeren AAM (1993) Nuclear DNA synthesis during the induction of embryogenesis in cultured microspores and pollen of Brassica napus L. Theor Appl Genet 87:9 – 16

Blomstedt CK, Knox RB, Singh MB (1996) Generative cells of Lilium longiflorum possess translatable mRNA and functional protein synthesis machinery. Plant Mol Biol 31:1083 – 1086

Bouget F-Y, Gerttula S, Shaw S, Quatrano RS (1996) Localization of actin mRNA during the establishment of cell polarity and early cell divisions in Fucus embryos. Plant Cell 8:189 – 201.

Brewbaker JL (1967) The distribution and phylogenetic significance of binucleate and trinucleate pollen grains in the Angiosperms. Amer J Bot 54:1069 – 1083

Brink RA, MacGillivray JH (1924) Segregation for the waxy character in maize pollen and differential development of the male gametophyte. Am J Bot 11:465 – 469

Brown SM, Crouch ML (1990) Characterization of a gene family abundantly expressed in Oenothera organensis pollen that shows sequence similarity to polygalacturonase. Plant Cell 2:263 – 274

Brown RC, Lemmon BE (1991a) Pollen development in orchids. 1. Cytoskeleton and the control of division plane in irregular patterns of cytokinesis. Protoplasma 163:9 – 18

Brown RC, Lemmon BE (1991b) Pollen development in orchids. 2. The cytokinetic apparatus in simultaneous cytokinesis. Protoplasma 165:155 – 166

Brown RC, Lemmon BE (1991c) Pollen development in orchids. 3. A novel generative pole microtubule system predicts unequal pollen mitosis. J Cell Sci 99:273 – 281

Brown RC, Lemmon BE (1991d) Pollen development in orchids. 5. A generative cell domain involved spatial control of the hemispherical cell plate. J Cell Sci 100:559 – 565

Brown RC, Lemmon BE (1992). Pollen development in orchids. 4. Cytoskeleton and ultrastructure of the unequal pollen mitosis in Phalaenopsis. Protoplasma 167:183 – 192

Brumfield R. (1941) Asymmetric spindles in the first microspore division of certain angiosperms. Amer J Bot 28:713 – 722

Cai G, Moscatelli A, Del Casino C, Cresti M (1996) Cytoplasmic motors and pollen tube growth. Sex Plant Reprod 9:59 – 64

Cameron DR, Moav RM (1957) Inheritance in Nicotiana tabacum. XXVII. Pollen killer, an alien genetic locus inducing abortion of microspores not carrying it. Genetics 42:326 – 335

Carmichael JS, Friedman WE (1995) Double fertilization in Gnetum gnemon – the relationship between the cell-cycle and sexual reproduction. Plant Cell 7:1975 – 1988

Chaudhury AM (1993) Nuclear genes controlling male fertility. Plant Cell 5:1277 – 1283

Chen Y-C, McCormick S (1996) Sidecar pollen, an Arabidopsis thaliana male gametophytic mutant with aberrant cell divisions during pollen development. Development 122:3243 – 3253

Di Laurenzio L, Wysocka-Dille, J, Malamy J.E, Pysh L, Helariutta Y, Freshour G, Hahn MG, Feldmann KA, Benfey .N (1996) The SCARECROW gene regulates an asymmetric cell division that is essential for generating the radial organization of the Arabidopsis root. Cell 86: 423 – 433

Doll H (1967) Segregation frequencies of induced chlorophyll mutants in barley. Hereditas 58:464 – 472

Dupuis I, Pace GM (1993) Factors affecting in vitro maturation of isolated maize microspores Plant Cell Rep 12:564 – 568

Eady C, Lindsey K, Twell D (1994) Differential activation and conserved vegetative cell-specific activity of a late pollen promoter in species with bi- and tricellular pollen. Plant J 5:543 – 550

Eady C, Lindsey K, Twell D (1995) The significance of microspore division and division symmetry for vegetative cell-specific transcription and generative cell differentiation. Plant Cell 7:65 – 74

Errampalli D, Patton D, Castle L, Mickelson L, Hansen K, Schnall J, Feldman K, Meinke D (1991) Embryonic lethals and T-DNA insertional mutagenesis in Arabidopsis. Plant Cell 3:149 – 157

Evans TC, Crittenden SL, Kodoyianni V, Kimble J (1994) Translational control of maternal glp-1 mRNA establishes an asymmetry in the C. elegans embryo. Cell 77:183 – 194

Feldman KA (1991) T-DNA insertion mutagenesis in Arabidopsis: a mutational spectrum. Plant J 1:71 – 82

Ferrandon D, Elphick L, Nusslein-Volhard C, St Johnston D (1994) Staufen protein associates with the 3'UTR of bicoid mRNA to form particles that move in a microtubule-dependent manner. Cell 79:1221 – 1232

Freeling M (1976) Intragenic recombination methods in maize: pollen analysis methods and the effect of parental Adh+ isoalleles. Genetics 83:701 – 717

Geitler L (1935) Beobachtungen über dies erste teiung im pollenkorn der angiospermen. Planta 24:361 – 386

Gervais C, Simmonds DH, Newcomb W (1994) Actin microfilament organization during pollen development of Brassica napus cv. Topas. Protoplasma 183:67 – 76

Glotzer M, Hyman AA (1995) The importance of being polar. Curr. Biol. 5, 11021105

Guo S, Kemphues KJ (1995) par-1 a gene required for establishing polarity in C. elegans embryos, encodes a putative Ser/The kinase that is asymmetrically distributed. Cell 81:611 – 620

Hanson DD, Hamilton DA, Travis JI, Bashe DM, Mascharenhas JP (1989) Characterization of a pollen-specific cDNA clone from Zea mays and its expression. Plant Cell 1:173 – 179

Haskell DW, Rogers OM (1985) RNA synthesis by vegetative and sperm nuclei of trinucleate pollen. Cytologia 50:805 – 809

Hause G, Hause B, van Lammeren AAM (1991) Microtubular and actin-filament configurations during microspore and pollen development in Brassica napus L. cv. Topas. Can J Bot 70:1369 – 1376

Havlicky T, Hause G, Hause B, Pechan P, van Lammeren AAM (1996) Subcellular localization of cell-specific gene expression in Brassica napus L. and Arabidopsis thaliana L. Microspores and pollen by non-radioactive in situ hybridization. PhD thesis, G Hause and B Hause, Wageningen Argricultutal University, ISBN 90-5485-614-9

Heslop-Harrison J (1968) Synchronous pollen mitosis and the formation of the generative cell in massulate orchids. J Cell Sci 3:457 – 475

Heslop-Harrison J (1971) Wall pattern formation in angiosperm microsporogenesis. Symp Soc Exp Biol 25:277 – 300

Horvitz RH, Herskowitz I (1992) Mechanisms of asymmetric cell division: Two Bs or not two Bs, that is the question. Cell 68:237 – 255

Hulskamp M, Parekh NS Grini, Scheitz K, Zimmermann I, Lolle SJ, Pruitt RE (1997) The STUD gene is required for male-specific cytokinesis after telophase II of meiosis in Arabidopsis thaliana. Dev Biol 187:114 – 124

Jalouzot R (1969) Nuclear and cytoplasmic differentiation in the pollen grain of Lilium candidum. Ex Cell Res 55:1 – 8 (French)

Jones DF (1928) "Selective fertilization." University of Chicago Press, Chicago

Katsuta J, Hashiguchi Y, Shibaoka H (1990) The role of the cytoskeleton in positioning of the nucleus in premitotic tobacco BY-2 cells. J Cell Sci 95:413 – 422

Kindinger B, Beckett JB, Coe EH (1991) Differential effects of specific chromosomal deficiencies on this development of the maize pollen grain. Genome 34:579 – 594

Kloc M, Etkin LD (1995) Two distinct pathways for the localization of RNAs at the vegetal cortex in Xenopus oocytes. Development 121:287 – 297

Konieczny A, Ausubel FM 1993. A procedure for mapping Arabidopsis mutations using co-dominant ectotype-specific PCR-based markers. Plant J 4:403 – 410

Kraut R, Chia W, Jan LY, Jan YN, Knoblich JA (1996) Role of inscuteable in orientating asymmetric cell divisions in Drosophila. Nature 383:50 – 55

Kyo M, Harada H (1986) Control of the developmental pathway of tobacco pollen in vitro. Planta 168:427 – 432

LaCour LF (1949) Nuclear differentiation in the pollen grain. Heredity 3:319 – 338

Leyser O, Chang C (1996) In G. Foster and D. Twell (eds.) Plant Gene Isolation: principles and practice. John Wiley & Sons Limited, Chichester. pp247 – 271
Liu Y-G, Mitsukawa N, Oosumi T, Whittier RF (1995) Efficient isolation and mapping of Arabidopsis thaliana T-DNA insert junctions by thermal asymmetric interlaced PCR. Plant J 8:457 – 463
Loegering WQ, Sears ER (1963) Distorted inheritance of stem-rust resistance of timstein wheat caused by a pollen-killing gene. Can J Genet Cytol 5:65 – 72
Lukowitz W, Mayer U, Jurgens G (1996) Cytokinesis in the Arabidopsis embryo involves the syntaxin-related KNOLLE gene product. Cell 84:61 – 71
Mascarenhas JP (1966) Pollen tube growth and ribonucleic acid synthesis by vegetative and generative nuclei of Tradescantia. Am J Bot 53:563 – 569
Mascarenhas JP (1975) The biochemistry of angiosperm pollen development. Bot. Rev. 41:259 – 314
Mascarenhas JP (1992) Pollen gene expression. In: Russell SD, Dumas C (eds) International Review of Cytology: Sexual Reproduction in Flowering Plants, Academic Press, San Diego, pp3 – 18
Mayer U, Torres Ruiz RA, Berleth T, Misera S, Jurgens G (1991) Mutations affecting body organization in the Arabidopsis embryo. Nature 353:402 – 407
Mayer U, Buttner G, Jurgens G (1993) Apical-basal pattern formation in the Arabidopsis embryo: studies on the role of the gnom gene. Development 117:149 – 162
McCormick S (1993) Male gametophyte development. Plant Cell 5, 1265 – 1275
McCormick S, Chen S, Curie C, Eyal Y, Kulikauska, R (1994) Molecular and genetic analysis of pollen development. In: Stephenson AG, Kao T-H (eds) Pollen-pistil interactions and pollen tube growth. Curr Topics Plant Phys, ASPP Series 12:29 – 38
McCoy TJ, Smith LY (1983) Genetics, cytology, and crossing behaviuor of an alfalfa (Medicago sativa) mutant resulting in failure of the postmeiotic cytokinesis. Can J Genet Cytol 25:390 – 397
Meinke DW (1982) Embryo-lethal mutants of Arabidopsis thaliana: evidence for gametophytic expression of the mutant genes. Theor Appl Genet 63:381 – 386
Meinke DW, Baus AD (1986) Gametophytic gene expression in embryo-lethal mutants of Arabidopsis thaliana. In: Mulcahy DL, Bergamini G, Ottaviano E (eds) Biotechnology and ecology of pollen. Springer-Verlag, Berlin, New York
Mineyuki Y, Furuya M (1986) Involvement of a colchicine-sensitive cytoplasmic element in premitotic nuclear positioing of Adiantum protonema. Protoplasma 130:83 – 90
Moscatelli A, Del Casino C, Lozzi L, Cai G, Scali M, Tiezzi A, Cresti M (1995) High molecular weight polypeptides related to dynein heavy chains in Nicotiana tabacum pollen tubes. J Cell Sci 108:1117 – 1125
Moscatelli A, Cai G, Liu G-Q, Tiezzi A, Cresti M (1996) Dynein-related polypeptides in pollen and pollen tubes. Sex Plant Reprod 9:312 – 317
Muschietti J, Dircks l, Vancanneyt G, McCormick S (1994) LAT52 protein is essential for tomato pollen development: pollen expressing antisense lat52 RNA hydrates and germinates abnormally and cannot achieve fertilization. Plant J 6:321 – 338
Olmedilla A, Schrauwen JAM, Wullems GJ (1991) Visualization of starch-synthase expression by in situ hybridization during pollen development. Planta 184:182 – 186
Ottaviano E, Mulcah, DL (1989) Genetics of angiosperm pollen. Adv Genet 26:1 – 64
Ottaviano E, Petroni D, Pe ME (1987) Gametophytic expression of genes controlling endosperm development in maize. Theor Appl Genet 75:252 – 258

Owen HA, Makaroff CA (1995) Ultrastructure of microsporogenesis and microgametogenesis in Arabidopsis thaliana (L.) Heynh. ecotype Wassilewskija (Brassicaceac). Protoplasma 185:7 – 21

Pacini E (1990) Tapetum and microspore function. In: Blackmore S, Knox RB (eds) Microspores: Evolution and Ontogeny. Academic Press Ltd, London, pp213 – 237

Palevitz BA (1982) The stomatal complex as a model of cytoskeletal participation in cell differentiation. In: Lloyd CW (ed) The Cytoskeleton in Plant Growth and Development. Academic Press, London, pp345 – 376

Parnell FR (1921) Note on the detection of segregation by examination of the pollen of rice. J Genet 11:209 – 212

Poccia DL, Green GR (1992) Packaging and unpackaging the sea urchin sperm genome. TIBS (June 1992) 223 – 227

Pokrywka NJ, Stephenson EC (1995) Microtubules are a general component of mRNA localization systems in Drosophila oocytes. Dev Biol 167:363 – 370

Preuss D, Rhee SY, Davis RW (1994) Tetrad analysis possible in Arabidopsis with mutation of the QUARTET (QRT) genes. Science 264:1458 – 1460

Raghavan V (1989) mRNAs and a cloned histone gene are differentially expressed during anther and pollen development in rice (Oryza sativa L.). J Cell Sci 92:217 – 229

Redei G P (1965) Non-Mendelian megagametogenesis in Arabidopsis. Genetics 51:857 – 872

Regan S, Moffatt BA (1990). Cytochemical analysis of pollen development in wild-type Arabidopsis and a male sterile mutant. Plant Cell 2:877 – 889

Reijnen WH, van Herpen MMA, de Groot PFM, Olmedilla A, Schrauwen JAM, Weterings KAP, Wullems GJ (1991) Cellular localization of a pollen-specific mRNA by in situ hybridization and confocal laser scanning microscopy. Sex Plant Reprod 4:254 – 257

Reynolds TL, Raghavan V (1982) An autoradiographic study of RNA synthesis during maturation and germination of pollen grains of Hyoscyamus niger. Protoplasma 111:177 – 188

Rhyu MS, Knoblich JA (1995) Spindle orientation and asymmetric cell fate. Cell 82:523 – 526

Roberts M (1996) Transposon-mediated insertional mutagenesis. In Foster G, Twell D (eds) Plant Gene Isolation: principles and practice. John Wiley & Sons Limited, Chichester, pp301 – 328

Sari Gorla M, Rovida E (1980) Competitive ability of maize pollen. Intergametophytic effects. Theor Appl Genet 57:37 – 41

Sari-Gorla M., Ferrario S, Villa M, Pe ME (1996) GAMS-1 - a gametophytic male-sterile mutant in maize. Sex Plant Reprod 9:216 – 220

Sautcr JJ (1969) Cytochcmical invcstigations on histoncs in cclls diffcring in RNA and protein synthesis. Z Pflanzenphysiol 60:434 – 449 (German)

Schopfer CR, Hepler PK (1991) Distribution of membranes and the cytoskeleton during cell plate formation in pollen mother cells of Tradescantia. J Cell Sci 100:717 – 728

Schrauwen JAM, de Groot PFM, van Herpen MMA, van der Lee T, Reynen WH, Weterings KAP, Wullems GJ (1990) Stage-related expression of mRNAs during pollen development in lily and tobacco. Planta 182:298 – 304

Shen JB, Hsu FC (1992) Brassica anther-specific genes: characterization and in situ localization of expression. Mol Gen Genet 234:379 – 389

Sheridan WF (1973) Nonaqueous isolation of nuclei from lily pollen and an examination of their histones. Z Pflanzenphysiol Bd 68S450 – 459

Shevell DE, Leu W-M, Gilimor CS, Xia G, Feldman KA, Chua N-H (1994) EMB30 is essential for normal cell division, cell expansion, and cell adhesion in Arabidopsis and encodes a protein that has similarlity to Sec7. Cell 77:1051 – 1062

Singh MB, O'Neil PM, Knox RB (1985) Initiation of postmeotic ß-galactosidase synthesis during microsporogenesis in oilseed rape. Plant Physiol 77:225 – 228

Singleton WR, Mangelsdorf PC (1940) Gametic lethals on the fourth chromosome of maize. Genetics 25:366 – 390

Smyth DR (1994) Pollen clusters. Curr Biol 4:851 – 853

Spielman ML, Preuss D, Li F-L, Browne WE, Scott R, Dickinson HG (1997) TETRASPORE is required for male meiotic cytokinesis in Arabidopsis thaliana. Development 124:2645 – 2657

Sprague GF (1933) Pollen tube establishment and deficiency of waxy seeds in certain maize crosses. Proc Natl Acad Sci USA 19:838 – 841

Springer P S, McCombie W R, Sundaresan V, Martienssen R A (1995) Gene trap tagging of PROLIFERA, an essential MCM2-3-5-like gene in Arabidopsis. Science 268:877 – 880

Stadler LJ (1933) On the genetic nature of induced mutations in plants. II. A haplo-viable deficiency in maize. Res Bull Mz Agric Exp Stn 204:29

Stadler LJ, Roman H (1948) The effect of X-rays upon mutation of a gene A in maize. Genetics 33:273 – 303

Stinson J, Mascarenhas JP (1985) Onset of alcohol dehydrogenase synthesis during microsporogenesis in maize. Plant Physiol 77:222 – 224

Stoppin V, Vantard M, Schmit A-C, Lambert A-M (1994) Isolated plant nuclei nucleate microtubule assembly: the nuclaer surface in higher plants has centrosome-like activity. Plant Cell 6:1099 – 1106

Sundaresan V, Springer P, Volpe T, Haward S, Jones JDG, Dean C, Ma H, Martienssen R (1995) Patterns of gene action in plant development revealed by enhancer trap and gene trap transposable elements. Genes Dev 9:1797 – 1810

Sunderland N, Huang B (1987) Ultrastructural aspects of pollen dimorphism. In: Giles KL, Prakash J (eds) International Review of Cytology. Pollen: Cytology and Development, Academic Press Inc, London, pp175 – 220

Tanaka I, Ito M (1980) Induction of typical cell division in isolated microspores of Lilium longiflorum and Tulipa gesneriana. Pl Sci Lett 17:279 – 285

Tanaka I, Ito M (1981) Control of division patterns in explanted microspores of Tulipa gesneriana. Protoplasma 108:329 – 340

Terasaka O (1982) Nuclear differentiation of male gametophytes in gymnosperms. Cytologia 47:27 – 46

Terasaka O, Niits, T (1987) Unequal cell division and chromatin differentiation in pollen grain cells. I. centrifugal, cold and caffeine treatments. Bot Mag Tokyo 100:205 – 216

Terasaka O, Niitsu T (1990) Unequal cell division and chromatin differentiation in pollen grain cells. II. Microtubule dynamics associated with the unequal cell division. Bot Mag Tokyo 103:133 – 142

Terasaka O, Niitsu T (1995) The mitotic apparatus during unequal microspore division observed by a confocal laser scanning microscope. Protoplasma 189:187 – 193

Terasaka O, Tanaka R (1974) Cytological studies on the nuclear differentiation in microspore division of some angiosperms. Bot Mag Tokyo 87:209 – 217

Theerakulpisut P, Xu H, Singh MB, Pettitt JM, Knox RB (1991) Isolation and developmental expression of Bcp1, an anther-specific cDNA clone in Brassica campestris. Plant Cell 3:1073 – 1084

Thomas C (1996) PCR techniques. In: Foster G, Twell D (eds) Plant Gene Isolation: principles and practice. John Wiley & Sons Limited, Chichester, pp331 – 368

Tiezzi A, Moscatelli A, Cai G, Bartalesi A, Cresti M (1992) An immunoreactive homolog of mammalian kinesis in Nicotiana pollen tubes. Cell Motil Cytoskel 21:132 – 137

Torres Ruiz RA, Jurgens G (1994) Mutations in the FASS gene uncouple pattern formation and morphogenesis in Arabidopsis development. Development 120:2967 – 2978

Touraev A, Lezin F, Heberle-Bors E, Vicente O (1995) Maintenance of gametophytic development after symmetrical division in tobacco microspore culture. Sex Plant Reprod 8:70 – 76

Tupy J, Rihova L, Zarsky V (1991) Production of fertile tobacco pollen from microspores in suspension culture and its storage for in situ pollination. Sex Plant Reprod 4:284 – 287

Twell D (1992) Use of a nuclear-targeted ß-glucuronidase fusion protein to demonstrate vegetative cell-specific gene expression in developing pollen. Plant J 2:887 – 892

Twell D (1994) The diversity and regulation of gene expression in the pathway of male gametophyte development. In: Scott RJ, Stead AD (eds) Molecular and Cellular Aspects of Plant Reproduction, (Society for Experimental Biology, Cambridge, United Kingdom: Cambridge University Press), Seminar series 55, pp83 – 135

Twell D, Patel S, Sorensen A, Roberts M, Scott R, Draper J, Foster G (1993) Activation and developmental regulation of an Arabidopsis anther-specific promoter in microspores and pollen of Nicotiana tabacum. Sex Plant Reprod 6:217 – 224

Twell D, Yamaguchi J, McCormick S (1990) Pollen-specific gene expression in transgenic plants: Coordinate regulation of two different tomato gene promoters during microsporogenesis. Development 109:705 – 713

Ueda K, Tanaka I (1995a) Male gametic nucleus-specific H2B and H3 histones. Designated gH2B and gH3, in Lilium longiflorum. Planta 197:289 – 295

Ueda K, Tanaka I (1995b) The appearance of male gamete-specific histones gH2B and gH3 during pollen development in Lilium longiflorum. Dev Biol 169:210 – 217

Ursin VM, Yamagushi J, McCormick S (1989) Gametophytic and sporophytic expression of anther-specific genes in developing tomato anthers. Plant Cell 1:727 – 736

Van Lammeren AAM, Bednara J, Willemse MTM (1989) Organization of the actin cytoskeleton in Gasteria verrucosa (Mill.) H. Duval visualized with rhodamine-phalloidin. Planta 178:531 – 539

Vizir IY, Anderson ML, Wilson ZA, Mulligan BJ (1994) Isolation of deficiencies in the Arabidopsis genome by g-irradiation of pollen. Genetics 137:1111 – 1119

Weterings KA. (1994) Pollen gene regulation. Ph.D. dissertation, University of Nijmegen. ISBN 90-9006767-1

Xu H, Knox RB, Taylor PE, Singh MB (1995) Bcp1 a gene required for male fertility in Arabidopsis. Proc Natl Acad Sci USA 92:2106 – 2110

Zaki MAM, Dickinson HG (1991) Microspore-derived embryos in Brassica: the significance of division asymmetry in pollen mitosis I to embryogenic development. Sex Plant Reprod 4:48 – 55

Zhao Z-Y, Weber DF (1989) Male gametophyte development in monosomics of maize. Genome 32:155 – 164

Zhang G, Gifford DJ, Cass DD (1993) RNA and protein synthesis in sperm cells isolated from Zea mays L. Sex Plant Reprod 6:239 – 243

Zhang D, Nicklas RB (1996) "Anaphase" and cytokinesis in the absence of chromosomes. Nature 382:466 – 468

7 Use of Haplo-Diploidisation for Plant Breeding

G. Pelletier
Station de Génitique et d'Amélioration des Plantes, INRA Centre de Versailles, 78026 Versailles Cedex, France

7.1 Introduction

Before the discovery by Guha and Maheshwari (1964,1966) and Bourgin and Nitsch (1967) of in vitro androgenesis, the presence of haploids in sexual offsprings had been described in a large number of species (Kimber and Riley, 1963) but their use in plant breeding was limited. For instance, maize (Chase, 1969) potato (Hougas et al., 1964) and asparagus (Thévenin, 1968) were the object of an intensive and systematic production of haploids for breeding but only in some research teams. These teams invested in this method, despite the very low frequency of haploids among the seeds, the constraints of the methods for inducing and/or screening haploids and the difficulties of chromosome doubling by colchicine. These limiting factors drastically reduced the interest of haploids in plant breeding as substitutes to the normal inbreeding methods for obtaining homozygosity (self pollination or full-sib crosses).

The number of pollen grains produced by a plant being in large excess compared to the number of seeds, it appeared clearly that in vitro androgenesis, when generalized to crop species, could become the ideal method: even with a low frequency of haploid production per plated anther, the absolute number of haploid plants could be greatly superior to the number of haploids produced in situ, spontaneously or after induction.

At the beginning of the seventies an important research effort was carried out, often successful, to generalize the in vitro culture of anther or microspores to numerous species , to find alternative strategies using in vitro culture of female organs and finally, to improve the techniques, previously described, of in situ induction of haploids (for reviews on methods see Jain et al., Eds., 1996).

7.2 Why do we want to use haploids in plant breeding?

The central criterium governing the use of haplodiploidisation in selection is the cost of production of a doubled haploid (DH) and agronomic evaluation of the derived line. This cost has to be compared to the production and the evaluation of an equivalent line obtained by the inbreeding system which is available in the species concerned. These costs must take into account the time which is necessary for these operations in each case.

Figure 7.1 illustrates the fact that according to the constraints on obtaining doubled haploids and the level of complexity of the character which is being considered haplodiploidisation will be or not of interest.

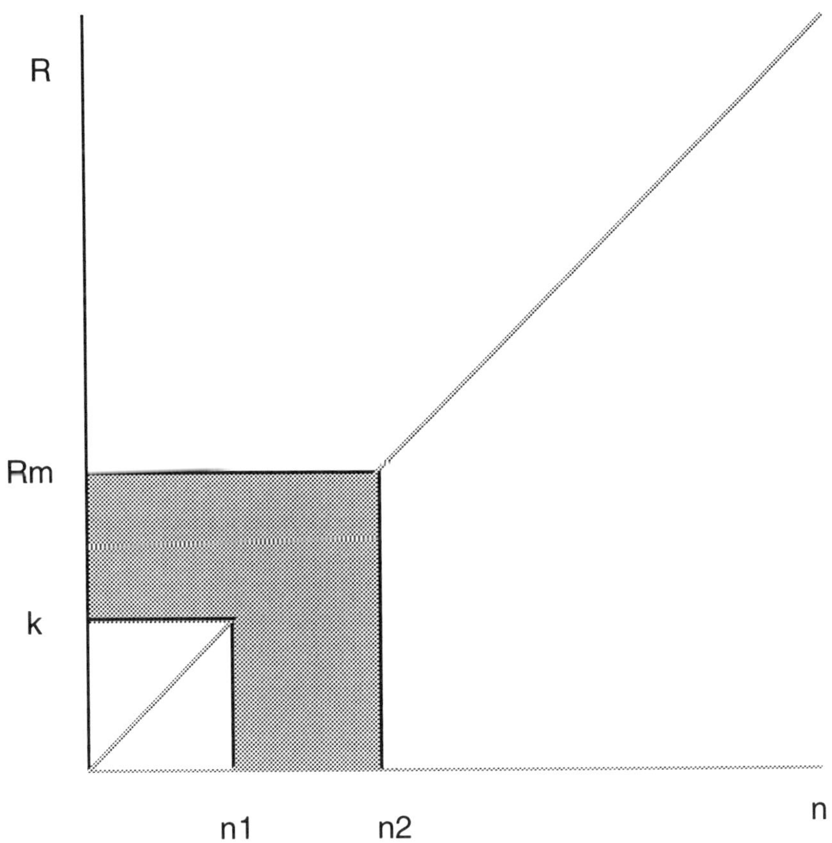

Fig. 7.1. Interval determined by complexities (n1 and n2) of the selected characters in which haplodiploidisation is of interest.

First, it is evident that one must produce more individuals through sexual reproduction (D) than through haplodiploidisation (DH) to have the same chance of obtaining a combination of "n genetic factors" (such as alleles, loci, QTLs, chromosomes,...) in one line. R, the ratio of these numbers (D/DH) is therefore superior to 1. As an example, if we consider doubled haploid lines derived from an F1 hybrid and homozygous lines in F2, R is an exponential function of the number of factors, n.

However, it is in general less expensive to produce an individual by sexual reproduction than by haplodiploidisation: the ratio k = "cost HD/cost D" is superior to 1. The k value decreases if we consider species with very long generation times (perennial), and is near zero if sexual reproduction is absent (apomixis). The haplodiploidisation method is useful only when R is superior to k. Finally, every breeding program has a maximum economic cost from which one can deduce the maximum quantity of work and therefore of plant material which can be produced and studied. One can consider that the ratio R cannot exceed a limit value, Rm, which is the consequence of these practical limitations. Rm and k define an interval of complexity corresponding to n1 and n2 in which it is advantageous to use the haplodiploidisation method.

7.3 Examples of the use of haploids for breeding in some species

Figure 7.2 illustrates three groups of species in which doubled haploids are used, according to the ease with which these doubled haploids can be produced. This factor determines the various uses of DH lines which have been made.

7.3.1 First group: doubled haploids are reproducibly produced but with a relatively low efficiency.

Rubber (Hevea brasiliensis) plantations which cover several million hectares in South-East Asia, are derived from a few individuals imported a hundred years ago from Brazil. Breeding and selection of this perennial, tetraploid and allogamous species is practically non-existent, but research at The Institute of Genetics in Beijing have resulted in reproducible techniques for producing haploids by anther cultures (Chen 1990). The pure lines which can be derived from these haploids after chromosome doubling are a first step towards a rationalization of breeding work in this species. The case of poplar, another woody species, could be cited as well (Wu and Nagarajan 1990).

In maize, Chinese workers pioneered the use of in vitro androgenesis, working with populations responding particularly well to this technique (Anonymous 1975). As early as 1985, the hybrid 411 x Qun Hua, in which the second parent is a DH line , was grown (Wu 1986). The major breeding firms in the US have

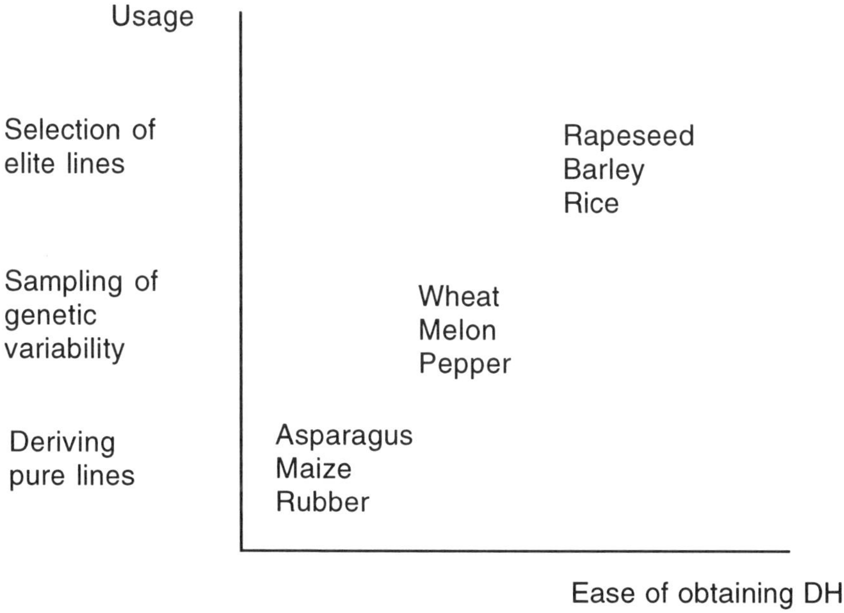

Fig. 7.2. Usage according to the ease of obtaining DH lines.

integrated the production of DH lines in their breeding programs: such lines are included in combination tests in the same way as pure lines derived from inbreeding (Petolino and Jones 1986).

The case of asparagus is a good example of the impact of haplodiploidisation on the breeding strategy: the development of this technique has drastically modified the types of varieties which are now selected and in some case commercialized. This species is perennial and dioecious, and more than forty years would be necessary by full-sib crosses to reach homozygosity. To obtain pure lines, haplodiploidisation is of course the method of choice. Early on, gynogenetic haploids spontaneously produced by parthenogenesis were isolated from twin seedlings (Randall and Rick 1945). In vitro androgenesis (Pelletier et al. 1972) subsequently allowed the production of female (mm) as well as super-male (MM) doubled haploids from male (Mm) individuals (Doré 1974). A selection scheme, using homozygous female and supermale parents led to the production of 100% male F1 hybrid varieties which offer various agronomical advantages: higher productivity, earliness, better longevity, no volunteers from seeds, which are a consequence, in this species, of the male phenotype. The first F1 hybrid of this type, Andreas, was commercialized by INRA and its partners in 1990. The same scheme is at present followed by breeders in France and Italy (Doré 1990).

7.3.2 Second group: a relatively large number of doubled haploids can be produced

In these species, in vitro androgenesis (Sibi et al. 1979) and parthenogenesis (Pochard and Dumas de Vaulx 1979) in the case of pepper, pollen irradiation followed by in vitro embryo culture in the case of melon (Sauton and Dumas de Vaulx 1987), and in vitro androgenesis or interspecific crosses with maize followed by embryo culture (in the case of wheat) allow the production of enough doubled haploids to have a convenient representation of the genetic variability arising in the progeny of a cross between two lines.

Two types of applications are possible. Firstly, DH populations can be used to establish genetic maps of phenotypic or molecular markers; in this case DH derived from an F1 hybrid are substitutes for recombinant lines obtained by single seed descent after about ten generations of selfing.

Secondly, DH populations are very useful to split up complex characters such as polygenic disease resistances, if we consider that these lines, which are directly fixed at the homozygous state, allow:

- repetitions in order to have a greater precision for measuring quantitative characters.

- the exposure of the same segregating progeny to different pathogenic strains under different conditions (inoculation stage, environmental factors).

Very significant progress has been obtained in the case of pepper for resistance to several diseases, particularly CMV, thanks to this analytical approach (Caranta and Palloix 1996, Lefebvre et al 1995), with the production, though recurrent selection methods, of commercial varieties for the Mediterranean area since 1990. In China, 7 DH varieties and six haploid-derived F1 hybrids have been commercialized since 1982.

In the case of wheat, mapping studies make use of DH lines and during the last decade, DH varieties have been registered in France (Florin) and in China (de Buyser et al 1987, Hu Han 1986).

7.3.3 Third group: the production of DH lines has no limiting factors.

In the case of rice, in vitro androgenesis has been essentially experimented in the sub-species japonica. In 1985, Hu quoted 81 varieties or precommercial lines obtained through this method in China. The International Research Institute on Rice in Philippines manages a programme of systematic production of haploids for several countries using the sub-species indica. From 1989 to 1993, 292 crosses (F1) produced after the in vitro culture of about 900 000 anthers, gave 4143 DH lines which are in the course of evaluation (Alejar et al. 1995).

In the case of barley, the majority of breeding laboratories are using haplodiploidisation either through the so-called "bulbosum method" or through microspore culture according to whether the genotypes are spring or winter types respectively (Devaux 1992).

In the case of rapeseed, in vitro microspore culture is routinely used in pedigree selection programs (Henderson and Pauls 1992) and more recently for the production of parental lines of F1 hybrid varieties.

Table 7.1. Some varieties derived from haploid plants

Maize	1950 s	Dekalb 640	USA
Rapeseed	1970 s	Maris-Haplona	UK
Tobacco	1974	F 211	Japan
	1978	Tanyu 1, 2, 3	China
	1984	NC 744	USA
Rice	1978	Huayu1, 2	China
		Tanfong 1	
		Huapei 1	
		Lunghug 1	
	1985-90	> 8 varieties	
	1992-93	Aya	Japan
		Shirayukihimine	
		Hirohikari	
Pepper	1982-86	13 varieties	China
Barley	1980	Mingo	Canada
	1984	Rodeo	
	1985	Gwylam	
	1989	Michka	France
	1991	Vodka	
Wheat	1987	Florin	France
	1985-90	> 6 varieties	China
Asparagus	1990	Andreas	France
Egg-plant	1990-95	Megal,Vernal,Tasca Barbara, Adria	France

7.4 Conclusion

About thirty years after the discovery of in vitro androgenesis, we can recognize that it catalyzed abundant research in order to generalize and improve the various methods for producing haploids in a wide range of species. Two to three thousand papers have been published on this topic during this period. Very quick progress was obtained which allowed a rapid integration of the method into breeding programs and given the resulting shortening of selection cycles, the haplodiploidisation method produced commercial varieties as early as the end of the seventies . It is noteworthy that the country which invested the most intensively in this method, both by the number of species concerned as well as the quantity of haploids produced is without contest China. In a difficult context, beginning with a plant material containing few genetic improvements, due to a relatively low cost of production of haploids in China, this method has been a factor of mobilization and genetic progress.

References

Alejar MS, Zapata FJ, Senadhira D, Khush GS, Datta SK (1995) Utilization of anther culture as a breeding tool in rice improvement. In :Terzi M et al (eds) Current Issues in Plant Molecular and Cellular Biology. Kluwer Academic Publishers, Dordrecht, Boston, London, pp137–142

Anonymus, 401 Research Group (1975) Primary study on induction of pollen plants of Zea mays (Engl. Abstr.) Acta Genet Sin 2:143

Bourgin JP, Nitsch JP (1967) Obtention de Nicotiana haploides à partir d'étamines cultivées in vitro. Ann Physiol Veg 9:377–382

Caranta C, Palloix A (1996) Both common and specific genetic factors are involved in polygenic resistance of pepper to several potyviruses. Theor Appl Genet 92:15–20

Chase SS (1969) Monoploids and monoploid-derivatives of maize (Zea mays L.). Bot Rev 35:117–167

de Buyser J, Henry Y, Lonnet P, Hertzog R, Hespel A, (1987) "Florin": a doubled haploid wheat variety developed by the anther culture method. Plant Breed 98:53–56

Devaux P (1992) Haploidy in barley and wheat improvement. In : Dattée Y et al (eds) Reproductive Biology and Plant Breeding. Springer-Verlag, Berlin, pp139–151

Chen Z (1990) Rubber (Hevea brasiliensis Muell. Arg.): In vitro production of haploids. In: Bajaj YPS (ed.) Biotechnology in Agriculture and Forestry, Vol. 12. Haploids in Crop Improvement I. Springer-Verlag, Berlin, pp215–236.

Dore C (1974) Production de plantes homozygotes mâles et femelles à partir d'anthères d'asperge cultivées in vitro. C R Acad Sci 278(17):2135–2138

Dore C (1990) Asparagus anther culture and field trials of dihaploids and F1 hybrids. In: Bajaj YPS (ed.) Biotechnology in Agriculture and Forestry, Vol. 12. Haploids in Crop Improvement I. Springer-Verlag, Berlin, pp322–345

Guha S, Maheshwari SC (1964) In vitro production of embryos from anthers of Datura. Nature 204:497

Guha S. Maheshwari SC (1966) Cell division and differenciation of embryos in the pollen grains of Datura in vitro. Nature 212:97–98

Henderson CAP, Pauls KP (1992) The use of haploidy to develop plants that express several recessive traits using light seeded canola (Brassica napus) as an example. Theor Appl Genet 83:476–479

Hougas RW, Peloquin SJ, Gabert AC (1964) Effect of seed parent and pollinator on frequency of haploids in Solanum tuberosum. Crop Sci 4:593–595

Hu H (1985) Use of haploids in crop improvement. In: Biotechnology in International Agricultural Research, International Rice Research Institute, P.O. Box 933, Manila, Philippines, pp75–84

Hu H (1986) Wheat: improvement through anther culture. In: Bajaj YPS (ed) Biotechnology in Agriculture and Forestry, Vol 2 Crops I. Springer-Verlag, Berlin, pp55–72

Kimber G, Riley R (1963) Haploid angiosperms. Bot Rev 29:490–531

Jain MS, Sopory SK, Veilleux RE (eds) (1996) In vitro haploid production in higher plants. Vol.1, Fundamental aspects and methods. Kluwer Academic Publishers, Dordrecht, Boston, London. 356p

Lefebvre V, Palloix A, Caranta C, Pochard E (1995) Construction of an intraspecific integrated linkage map of pepper using molecular markers and doubled haploid progenies. Genome 38:112–121

Pelletier G, Raquin C, Simon G (1972) La culture in vitro d'anthères d'asperge (Asparagus officinalis). C R Acad Sci 274(6):848–851

Petolino JF, Jones AM (1986) Anther culture of elite genotypes of Zea mays L. Crop Sci 26:1072–1074

Pochard E, Dumas de Vaulx R (1979) Haploid parthenogenesis in Capsicum annuum L. The biology and taxonomy of the Solanaceae. Linnean Soc Symp Ser 7 pp455–472

Randall TE, Rick CM (1945) A cytogenetic study of polyembryony in Asparagus officinalis L. Am J Bot 32:560–569

Sauton A, Dumas de Vaulx R (1987) Obtention de plantes haploides chez le melon (Cucumis melo) par gynogenèse induite par du pollen irradié. Agronomie 7:141–148

Sibi M, Dumas de Vaulx R, Chambonnet B (1979) Obtention de plantes haploides par androgenèse in vitro chez le piment (Capsicum annuum L.). Ann Amel Plantes 29:583–606

Thevenin L (1968) Les problèmes d'amélioration chez Asparagus officinalis. II. Haploidie et amélioration. Ann Amel Plantes 18(4):327–365

Wu JL (1986) Breeding haploid corn by anther culture. In: Hu H, Yang HY (eds) Haploids of higher plants in vitro. Springer-Verlag, Berlin, pp149–161

Wu K., Nagarajan P (1990) Polars (Populus spp.): In vitro production of haploids. In: Bajaj YPS (ed) Biotechnology in Agriculture and Forestry, vol 12. Haploids in Crop Improvement I. Springer-Verlag, Berlin, pp237–249

8 What Are the Next Priorities for the Future?

Y. Henry
Laboratoire de Biologie du Développement des Plantes, IBP, ERS 569, Université Paris-Sud, 91405 Orsay, France

8.1 Introduction

The lectures of the IAPTC congress held in Orsay in the Spring of 1985 (De Buyser et al., 1986) on haploid technology revealed that highly significant improvements have been achieved over the last 10 years. To date, a great deal of effort has been devoted to the development and study of haploidy and thousands of papers have been published. It might therefore have been tempting to simply state, in conclusion, that most of the difficulties currently faced in our field will be appropriately resolved in due course. Indeed, such a conclusion is substantially borne out by recent progress in the field. Nevertheless, we know that 20 years from now, a large proportion of the current work will be of no more than historical interest. In 1996, working in the field of plant cell research is no longer a question of carefully and elegantly furnishing one's own isolated corner of the scientific labyrinth. On the contrary, it is important to open as many windows as possible, which is exactly what J.-P. Bourgin managed to do with great success. I would like to make special note of the fact that in the study of haploidy, the collective efforts of the various French research organizations (CNRS, INRA and Universities) have played an important pioneering role in the development of the technology involved. Moreover, much of this research was transferred to plant breeding companies, which developed their own breeding programs through this provision of a more rapid breeding method.

In this chapter, we will focus on both the present state of developments in the area of haploidy and new developments anticipated for future years. In addition, some insights into the use of this technology for cell and molecular biology purposes will be discussed. Even if the essential technology would appear to have been already developed or initiated, various interesting long term goals remain to be achieved and new developments can equally be expected to occur.

8.2 State of the art in 1996

The numerous *in situ* and *in vitro* technologies allowing the production of haploid and doubled haploid plants have been described in a previous chapter (Raquin & Horlow), demonstrating that many studies have been devoted to improving the efficiency of haploid technology. Although tissue culture conditions have been optimized for a wide range of plants, a number of problems remain. In several species we are still either unable to produce haploids or at best can obtain only a limited number of haploids (*eg* tomato, Arabidopsis and various tree species). Moreover, it has not yet been possible to achieve a uniformly high and genotype-independent tissue culture response for any plant (see the chapter from Beckert), even in the model species rape and tobacco.

On the other hand, the experimental work carried out on a number of species has led to various interesting observations, such as the appearance of plants of unexpected type. The *in vitro* culture of haploid microspores often produces embryos and plants which are polyploids (2n, 3n, 4n etc) or aneuploids. In most plants of the Poaceae family, albinos are often observed in microspore-derived plants, and albinism is considered to be the most frequent "gametoclonal variation" in these species. It has been observed in several species, mostly polyploid ones, that a limited number of doubled haploid lines produce heterogeneous and unstable progenies.

The various observations described, which are given as examples, lead one to ask how the observed "abnormalities" arise. It has been demonstrated that some of the observed abnormalities preexist in the microspores used for *in vitro* culture, notably those abnormalities associated with the heterozygosity of the starting material, even when inbred lines, including those showing disrupted meiotic activity, are used. Thus most of the abnormal meiotic behaviour in homozygous lines, even doubled haploid ones (De Buyser et al., 1985) prove to arise from the starting genotypes and explain most of the unstable doubled haploid lines observed in field experiments. On the contrary, the production of polyploid (2n, 3n, 4n etc) and aneuploid plants is induced by the *in vitro* culture. Endoreduplication of the microspore DNA happens early in the microspore nucleus, prior to the first embryogenic mitosis, and is induced by cold pretreatment and the tissue culture process itself (Raquin et al., 1982).

We can say that haploidisation is a remarkably powerful tool for all laboratories interested in crop improvement (see the chapter from Pelletier), since doubled haploids give rise to instant homozygous lines. Doubled haploids reached homozygosity in one step, and are thus of major interest for obtaining either homozygous transformed plants or homozygous dominant genes (*eg* Asparagus). In several examples, the use of doubled haploid technology allows rapid genetic progress to be made. This has been clearly illustrated in the case of the eggplant, for which the first microspore-derived haploids were produced in 1979 and the first F1 hybrid variety having a doubled haploid as a female parent was released in 1989.

Prior to the use of *in vitro* culture techniques, geneticists used spontaneous doubled haploids (*eg* for maize, potato and asparagus). The best example is that of Sears, who in the 1930s produced from a spontaneous haploid of wheat a large number of aneuploid lines (Figure 8.1) upon which studies of the genetics of bread wheat are founded to this day (Sears, 1939). Today, geneticists use the powerful tool of haploids and doubled haploids in a number of different ways. Stable lines

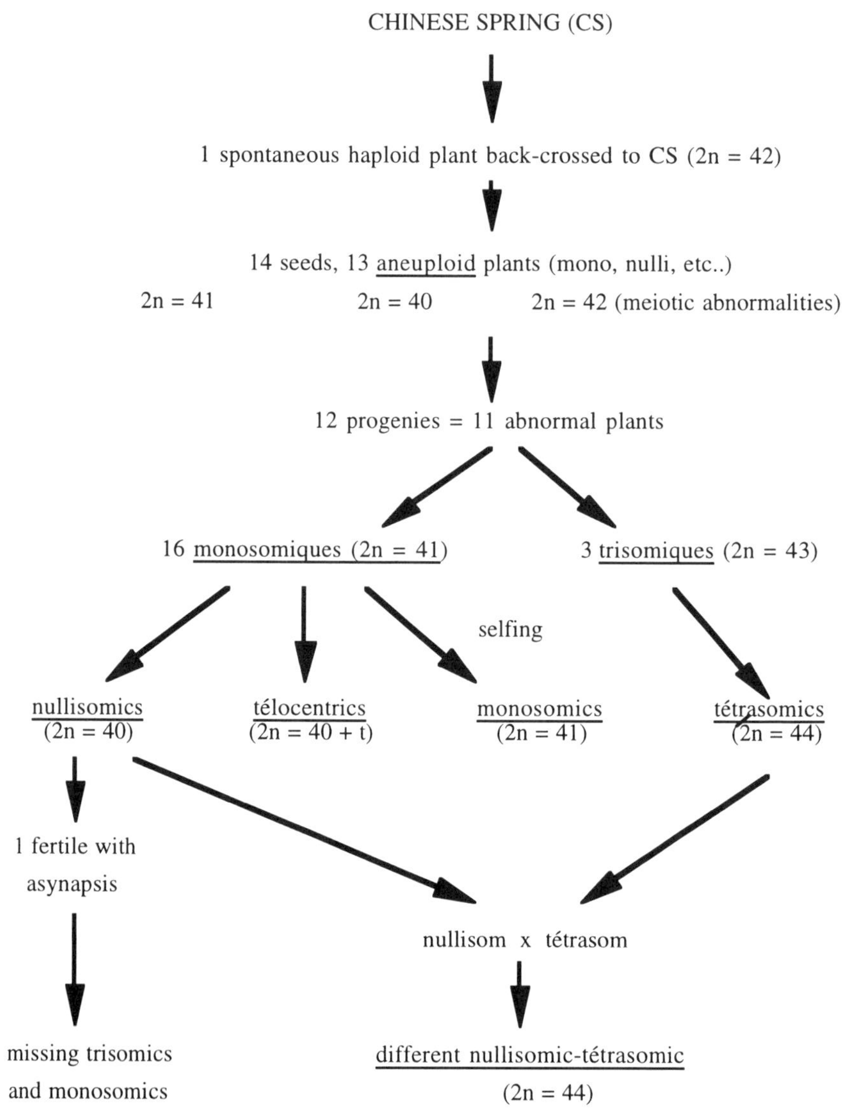

Fig. 8.1. Origin of most of the aneuploids in wheat (from Sears, 1939).

are necessary for genetic and mapping studies. Recombinant inbred (RI) lines were produced classically by selfing the progeny of individual F2 plants until achieving homozygosity, a process which requires several years. The use of doubled haploid (DH) lines, a particular subgroup of RI, reduces the time and simplifies the analysis. DH lines are therefore excellent tools for establishing genetic maps (*eg* wheat, barley, rape, sweet cherry, gymnosperms), for analyzing the genetic control of quantitative characters and for identifying QTLs. This opens up a wide field for future developments. As an example, doubled haploid wheat lines were produced in 1985 through interspecific crosses and anther culture of the F1 hybrid Sicco CS5B x Highbury CS 5B (Snape et al., 1986). These RI lines were successfully used to establish a physical mapping of the a-amylase-1 locus (Cheung et al., 1991), to develop extensive genetic mapping of the homoeologous group 7 chromosomes (Chao et al., 1991), to identify QTLs for major agronomic characteristics (Hyne et al., 1994) and to compare male and female meiotic recombination frequencies (Wang et al., 1995). Although unexpected deviant lines can be found, doubled haploid lines make genetic studies easier.

8.3 Priorities for the future

Those researchers who work on androgenesis have stated from the outset that the system provides an ideal model for the study of embryonic development. However the mechanisms involved in embryogenic induction are still poorly understood. Fortunately, in recent years, efficient *in vitro* systems for inducing embryogenesis in isolated microspores or young pollen grains have been established for rape and tobacco. My colleagues Heberle-Bors and Van Lockeren-Campagne have shown in preceding chapters that tobacco and rape microspore cultures are excellent model systems. In rape, microspores isolated shortly before the first pollen mitosis can be induced to form embryos by a 9 to 48 hour heat shock treatment at 30-32°C in a rich medium, followed by culture at 25°C (Lichter, 1982). The stress seems to be necessary as a trigger for both the overcome of the blocked cell cycle and the progress through the cell cycle. After the inductive stress, the first sporophytic mitosis of the microspore is followed by rapid cell divisions and embryogenesis. In tobacco, mid- to late-uninucleate microspores and early bicellular pollens cultured in a nutritionally poor (starvation) medium develop into embryogenic pollen grains (Kyo & Harada, 1986). The starvation treatment synchronize embryogenic material. After transfer to a "rich medium" (ie: containing sucrose), microspore and vegetative cells of the pollen grains divide and form embryos (Touraev et al., 1996). Within a wide developmental window, male gametophytes are susceptible to change their developmental program towards a sporophytic one. The initiation of embryogenesis seems to be due to the block of normal gametophytic development in response to a stress treatment. Thus for each of the two species, we have at our disposal an attractive source of biological material for studies of the cellular and molecular mechanisms involved in early embryogenesis. For example, it is conceivable that change from the normal male gametophyte development to microspore embryogenesis coincides with alterations in the configuration of microtubules and microfilaments (Hause & Hause, 1996).

Future studies of plant embryogenesis will thus include investigations of early signal transduction. The two stress signals, ie heat-shock and starvation, might produce similar intracellular effects. The rapid synthesis of signalling elements is a highly conserved cellular response, involving the redistribution of gene transcription and mRNA translation and thus playing a fundamental role in the regulation of development. The signalling cascade couples a large array of upstream events to a wide range of downstream physiological processes. We can expect that the study of plant signalling elements related to stress, to cell proliferation and to cell differentiation will be required for the study of microspore-derived embryogenesis.

In the same way, homogeneous populations of maize microspores have been purified in sufficient amounts for electrophoresis coupled to immunodetection an fluorography of *in vitro* synthesized proteins. This analysis has revealed that isolated microspores effect a classical heat-shock response characterized by the repression of protein synthesis and the expression of HSPs (Magnard et al., 1996).

In another aspect of the work presented here, the chapter presented by Twell illustrates the fact that numerous colleagues are interested in male gametophyte development and more precisely, in the functioning of haploid cell genomes. An interesting aspect which merits further study concerns the origin of instability in the chloroplast genome of cereal microspores. It may be hypothesized that certain molecular mechanisms already occur in the original plant material which are amplified by the culture procedure. Most of the albino plants obtained in this way have a heterogeneous population of ctDNA molecules (Day & Ellis, 1984), suggesting that the albinism originated from the degradation of ctDNA in proplastids during microsporogenesis. This heteroplasmy might be amplified during tissue culture. Either microspore ctDNA might be altered in several proplasts or some ctDNA molecules might be altered in all proplasts.

It is now possible to develop in one step alien chromosome addition lines using anther cultures of F1 hybrids between related species or genera. Everyone knows that the creation of alien chromosome addition lines in polyploid species by selfing is often practically impossible because of the very low frequency at which they occur (male gamete frequency x female gamete frequency). This problem might be easily overcome by using anther culture (Wang et al., 1993) ie, the male gametic frequency. Numerous applications will no doubt be developed for crop improvement in this way. An even more powerful use of this approach will be for the improvement of crop plants which no longer use the sexual mode of reproduction, as is the case for certain parthenocarpic varieties of banana. In this case the production of plants from meiotic products represented in the microspores might provide, paradoxically, a novel source of genetic variability.

Plant transformation technology has always been of great interest due to its potential benefits to crop improvement and subsequently to human nutrition. Whole chromosome, chromosome arm or chromosome fragment transfers from related species have been used (and will be used) for plant improvement for many years. Genetic engineering offers the opportunity to increase the number and types of useful transmissible traits. Direct DNA transfer into proliferative regenerable cells cultivated *in vitro* is an important step forward for genotype independent monocot transformation. An alternative to this strategy involves the biolistic transformation both of pre-reproductive cells *in situ* (Dupuis & Pace, 1993) and of regenerable microspores (Jähne et al., 1994). Microspores constitute a unicellular target, but a more interesting development in monocots and dicots should be the germline transformation. The use of pollen as a natural delivery vector for foreign DNA can provide an alternative means to transform plant species. This strategy requires the transfer of DNA through the pollen wall. Subsequently, the DNA must integrate into one of the chromosomes of the generative cells or the sperm cell. The transformed pollen grain might be used for pollination. Later on the sperm cells can deliver the foreign DNA to the egg cell. Although many aspects of this model are, at present, highly speculative, delivery of DNA into pollen by particle gun bombardment, and inheritance of the introduced DNA to the progeny has recently been demonstrated (Van der Leede-Plegt et al., 1995). A possible application of such a tool is the coupling of *in vitro* pollen maturation to *in vitro* selection (toxin, herbicides and so on), there is no doubt about the future developments.

8.4 Conclusions

In vitro cell and tissue culture was originally a biotechnological approach developed to assist plant breeding, but paradoxically, research concerned with haploidy has had two major consequences:

- The early successes of this work, which were often achieved in an empirical fashion, have led to some exceptional results being obtained with several different species (notably rape, tobacco and barley), making possible studies of the molecular mechanisms involved. Investigations of microspore embryogenesis has involved biochemical and molecular methods added by cytological and histological techniques. As a consequence, the EU is now funding several research programs involving signalling during early *in vitro* embryogenesis.
- Of the various different *in vitro* techniques currently used (*eg* protoplasts, somatic embryogenesis and transformation), haploid production and vegetative propagation has had the most direct applications in plant breeding. Compared to other *in vitro* technologies, the current scientific literature concerned with haploidy seems to be somewhat restricted. This is probably a consequence of

the high quality of the results obtained since 1964. In fact haploid technology has become what it should always be, aside from its biotechnological uses, namely a tool for cell biologists.

Haploid technology will clearly not replace conventional tools, but this plant cell biotechnology will benefit both molecular genetic studies and plant breeding.

Acknowledgments

I acknowledge the help of Jacques De Buyser, Aimé Nato, James Tregear and Philippe Vain.

References

Bourgin JP, Nitsch JP (1967) Obtention de Nicotiana haploïdes à partir d'étamines cultivées in vitro. Ann Physiol Vég 9:377 – 382

Chao S, Sharp PJ, Worland AJ, Warham EJ, Koener RMD, Gale MD (1989) RFLP-based genetic maps of wheat homoeologous groups 7 chromosomes. Theor Appl Genet 78:495 – 504

Cheung WI, Chao S, Gale MD (1991) Long-range physical mapping of the a-amylase-1 (a-Amy-1) loci on homoeologous group 6 chromosomes of wheat. Mol. Gene. Genet., 229, 373 – 379.

Day A, Ellis THN (1984) Chloroplast DNA deletions associated with wheat plant s regenerated from pollen : possible basis for maternal inheritance of chloroplasts. Cell 39:359 – 368

De Buyser J, Henry Y, Taleb G (1986) Wheat androgenesis: cytogenetical analysis and agronomic performance of doubled haploids. Z Pflanzenzuchtg 95:23 – 34

De Buyser J, Henry Y, Raquin C, Benbadis A, (1986) Obtention d'haploïdes in vitro – Etat actuel et perspectives.Société Botanique de France, Actualités Botaniques, 1986 – 4, Paris, 83p

Dupuis I, Pace GM (1993) Gene transfer to maize male reproductive structure by particle bombardment of tassel primordia. Plant Cell Reports, 12:607 – 611

Jähne A, Becker D, Brettschneider R, Lörz H (1994) Regeneration of transgenic, microspore-derived, fertile barley. Theo. Appl Genet 89:525 – 533

Lichter R (1982) Induction of haploid plants from isolated pollen of Brassica napus. Z. Pflanzenphysiol 105:427 – 434

Magnard JL, Vergne P, Dumas C (1996) Complexity and genetic variability of heat-shock protein expression in isolated maize microspores. Plant Physiol 11:1085 – 1096

Hause G, Hause B (1996) Induction of embryogenesis in isolated microspores and pollen of Brassica napus L. cv. Topas. PhD thesis, Wageningen, The Netherlands, November 19th 1996, 150p

Hyne ., Kearsey MJ, Martinez O, Gang ., Snape JW (1994) A partial genome assay for quantitative trait loci in wheat (Triticum aestivum) using different analytical techniques. Theor Appl Genet 89:735 – 741

Raquin C, Amssa M, Henry ., De Buyser J, Essad S (1982) Origine des plantes polyploides obtenues par culture d'anthères. Analyse cytophotométrique in situ et in vitro des microspores de Pétunia et de Blé tendre. Z Pflanzenzuchtg 89:265 – 277

Sears E (1939) Cytogenetic studies with polyploid species of wheat. I. Chromosomal aberrations in the progeny of a haploid of Triticum vulgare. Genetics 24:509 – 523

Snape JW, De Buyser J, Henry Y, Simpson E (1986) A comparison of methods of haploid production in a cross of wheat, Triticum aestivum. Z Pflanzenzuchtg 96:320 – 330

Touraev A, Pfosser M, Vicente O, Heberle-Bors E (1996) Stress as the major signal controlling the developmental fate of tobacco microspores: towards a unified model of induction of microspore/pollen embryogenesis. Planta 200:144 – 152

Van Der Leede-Plegt LM., Van De Veen BCE, Schilder M, Franken J, Van Tunen AJ (1995) Development of a pollen-mediated transformation method for Nicotiana glutinosa. Transgenic Research 4:77 – 86

Wang G, Ji J, Wang YB, Hu H, King IP, Snape JW (1993) The genetic characterization of a novel multi-addition doubled haploid lines derived from triticale x wheat hybrids. Theor Appl Genet 87:531 – 536

Wang G, Hyne V, Chao S, Henry Y, De Buyser J, Gale MD, Snape JW (1995) A comparison of male and female recombination frequency in wheat using RFLP maps of homoeologous group 6 and 7 chromosomes. Theor Appl Genet 91:744 – 746

Index of plants

Printing: Druckhaus Beltz, Hemsbach
Binding: Buchbinderei Schäffer, Grünstadt